Prompting the Future of Education

An Educator's Guide to AI Prompt Engineering

Dr. Eric A. Johnson

PeakEd Pathways

First printing, 2024.

Broken Table Press
Belchertown, MA, 01007

www.ericjohnsonwriter.com

To My Colleagues and Friends.

You are the ones who make

the daily challenges

of teaching manageable.

Prompting the Future of Education

Introduction

Over my twenty years in education, I've seen many changes come through my classroom, from technology shifts to evolving curricula. Each advancement has brought new opportunities and fresh challenges. Today, as AI emerges in our schools, educators face yet another shift, but one that promises to reshape how we teach, connect, and grow professionally. This book, *Prompting the Future of Education*, is my way of equipping teachers with the skills and insights needed to navigate this technological wave with confidence and creativity.

As a long-time proponent of educational technology, I was an early adopter, piloting one of the first Chromebook carts in my inner-city school and embracing the transition to digital resources. This journey ultimately prepared me for what may be the most transformative development yet: AI-powered tools like ChatGPT and Google Gemini, which offer the chance to streamline teaching tasks, customize learning experiences, and reduce some of the heavy workload that so often leads to burnout in our profession. However, like many teachers, I recognize the hesitation that surrounds new technology, especially when educators are already stretched thin with countless demands.

Yet, AI doesn't have to be another burden—it can be a powerful ally. Imagine a world where AI assists with lesson planning, grading, and individualized student support, allowing us to focus more on what truly matters: building connections, inspiring creativity, and guiding our students' personal growth. With AI, these possibilities are within reach, but only if we, as educators, take the time to understand how to work with it effectively. This book will teach you how to approach AI with confidence and use it to enhance—not replace—your role as an educator. Through the techniques and examples here, I want to show you that AI is more than a productivity tool; it's an extension of your own teaching strategy.

For AI to be genuinely transformative in education, we must go beyond surface-level use and develop skills in **prompt engineering**—the art of designing questions and instructions for AI in ways that yield rich, relevant, and practical results. I've structured this book to help you master this skill, offering real-world examples, clear strategies, and hands-on exercises to support you as you integrate AI thoughtfully into your classroom practices. From simplifying complex tasks to inspiring new instructional approaches, the prompts you create can make AI your classroom partner, responding to your specific needs with the nuance and adaptability that education demands.

This book is not only a guide but an invitation to reimagine how AI can benefit teaching and learning. It's written with the understanding that we have limited time, energy, and resources and that our work requires balance to be sustainable. By refining your approach to AI with well-crafted prompts, you'll be able to *offload repetitive tasks*, *maintain high-quality instruction*, and *reclaim precious time*. As you explore these chapters, I hope you will feel empowered to shape your relationship with AI into one that supports both your professional growth and your personal well-being. Integrating AI can enhance your teaching and simplify your day-to-day workload, giving you more space for creativity and reflection.

Each chapter of this book builds upon the last, taking you from foundational skills to advanced strategies. You'll discover how to adapt your prompts to achieve clarity, specificity, and flexibility, enabling AI to provide responses that are both informative and tailored to your educational goals. This journey will deepen your understanding of prompt engineering, transforming AI from a generalized tool into a customized classroom asset. By the end of this book, you'll have not only new skills but a refreshed outlook on what AI can offer in an educational setting, and (if I may be so bold) a renewed love for the profession to which we've dedicated out professional lives.

Today's generative AI tools are in their early stages, and now is the best time for us to adopt and adapt them. As AI technologies mature, they will become as integral to the educational landscape as slide decks, digital portals, and research databases. Choosing to learn and implement these skills now will position you ahead of the curve, benefiting not only your teaching but also the larger educational community. I encourage you to experiment with AI, join in this shift, and become a leader in modern education's digital transformation.

I've written *Prompting the Future of Education* with the understanding that each educator's journey with AI will be unique. The ideas, activities, and prompt examples provided are not meant to be prescriptive but to inspire and empower you to take this technology and make it your own. AI is the latest tool available to us, one that has the potential to reshape our teaching practices and provide the level of support and customization we've always hoped for. By approaching it thoughtfully, you can enrich your teaching, engage your students in new ways, and rediscover the joy that brought you to this profession. Together, we will take your prompts from this:

> Develop a lesson plan for a 10th-grade biology class covering the topic of photosynthesis.

to this:

> Use the attached dataset to discern the issues my class is having with standard 6.MS-LS1-2. Develop a lesson plan for a 10th-grade biology class covering the topic of photosynthesis. The plan should begin with a brief interactive lecture, include a lab experiment demonstrating the process using leaf discs in a baking soda solution under different light conditions, and conclude with a group discussion on the importance of photosynthesis in ecosystems. Provide step-by-step instructions and materials needed for the lab portion of the lesson.

Suddenly, you're seeing more relevant, specific results that save you planning time and bring richer, more engaging content to your classroom.

This shift from a basic request to a carefully tailored prompt is the essence of effective prompt engineering. It's about moving beyond the surface to craft instructions that unlock AI's full potential, turning it into a tool that adapts to your unique needs. When we go beyond basic prompts, we allow AI to truly understand and respond to the specifics needs in our individual classrooms.

This book isn't just about writing better prompts—it's about shifting our approach to technology in the classroom. With a bit of intention, we can go from passive users to active co-creators, using AI not just as a tool but as a partner in our teaching. By refining your prompt engineering, you're setting yourself up for a future where AI can support and enhance your teaching. This book is your roadmap to making that future a reality, helping you to lighten your workload, engage your students more fully, and rediscover your passion for teaching.

Packed with examples, suggested activities, and tested prompts, this guide demonstrates how precise prompts yield results that aren't just helpful but transformational. As educators, we're all stretched thin; by mastering prompt engineering, you can shift some of that load, creating space for what you love both inside and outside the classroom. This book is about thriving, not just surviving, in education.

Generative AI may be new, but it's here to stay. It will soon be as essential to our teaching as digital portals and research databases. By diving in now, you're setting the pace, becoming an early leader in using AI as an effective teaching partner, ready to embrace what's next.

Let's embark on this journey together. With each chapter, you'll gain the confidence, skills, and vision needed to incorporate AI thoughtfully into your practice. Together, we'll reshape education, making it more dynamic, flexible, and sustainable—now and in the years to come.

Chapter 1: Understanding the Basics of AI Language Models

Creativity follows mastery, so mastery of skills is the first priority for young talent.

~ Benjamin Bloom

A few months ago, a colleague came to me, completely exasperated. She'd been experimenting with ChatGPT but said, "I can't get it to do anything useful. It's like talking to a brick wall!" Curious, I asked her what kind of prompts she was using, and as she rattled off a few examples, it became clear they were overly rigid and lacked context. I suggested a simple shift: use a conversational tone—talk to the AI as if talking to a person. She tried, typing something like, "Hey, can you help me create a lesson plan for my ninth graders on verb forms?" The difference was immediate. The responses were detailed, relevant, and exactly what she needed. Her frustration turned to excitement: "I didn't realize it could be this helpful!" That moment underscored a vital truth: the way we interact with AI directly impacts the results we get.

In today's rapidly evolving educational landscape, Artificial Intelligence (AI) is no longer a futuristic concept; it's an essential tool reshaping how we teach and learn. AI language models like ChatGPT and Google Gemini have transformed educational technology, unlocking new ways to enhance instruction and engage learners. These tools, when used effectively, can help educators

> **AI is no longer a futuristic concept; it's an essential tool reshaping how we teach.**

streamline lesson planning, provide personalized learning experiences, and address long-standing challenges such as workload and diverse classroom needs. AI technology will play a pivotal role in shaping the future of teaching and learning, offering solutions that adapt to the unique needs of students while empowering teachers to focus on creative and interactive aspects of education.

The key lies in understanding how to communicate with these models effectively. The conversational approach isn't just a stylistic choice; it helps AI interpret intent, focus on context, and provide more meaningful, actionable responses. By integrating AI into teaching practices with clear, thoughtful prompts, educators can harness its potential to support everything from administrative tasks to creative lesson design.

AI's use in education isn't just about convenience—it's about transformation. With the ability to analyze vast amounts of data, AI tools can identify student learning patterns, suggest targeted interventions, and even adapt material to individual learning paces. This personalized approach helps address diverse learning needs while giving teachers more time to focus on what they do best: actually working with the students. However, like any tool, its effectiveness depends on how it's used. Learning to interact with AI properly is the first step in unlocking its transformative potential.

AI's Transformative Role in Education

AI is no longer just a buzzword—it's actively reshaping how we teach and learn by introducing tools that adapt to individual student needs, streamline administrative tasks, and enhance engagement. Imagine being able to pinpoint exactly where a student is struggling or having access to immediate, tailored solutions for their unique learning challenges. AI enables this level of customization, transforming traditional approaches to education into more flexible, personalized experiences.

Beyond these benefits, AI also empowers teachers by reducing their workload. Routine tasks like grading and lesson planning can now be handled more efficiently, freeing educators to focus on fostering creativity, critical thinking, and personal connections with their students. This shift isn't just about saving time —it's about allowing teachers to do what they do best: teach.

However, adopting AI isn't without its challenges. One significant concern is how students might misuse these tools, such as relying on AI to complete assignments without truly engaging with the material. While tempting, this approach can erode the development of vital skills like critical thinking and problem-solving. The solution lies in teaching students how to use AI responsibly. By framing AI as a collaborative tool—one that assists but doesn't replace effort—educators can guide students toward ethical and effective usage.

Another challenge is academic integrity. AI-generated content can blur the lines of originality, especially when students don't fully understand the importance of attribution. This makes it essential for schools to foster a culture of honesty and ethical awareness, helping students see AI as a resource to support learning rather than a shortcut.

While these challenges require thoughtful solutions, they shouldn't overshadow AI's transformative potential. For example, AI-driven tools can provide real-time feedback, recommend interventions for struggling learners, and adapt materials to different learning styles. In classrooms where students have diverse needs, AI offers the flexibility to ensure no one is left behind.

The key to unlocking these benefits is understanding how to interact with AI effectively. The quality of the results depends on the clarity and specificity of the prompts we provide. When educators refine their approach to communicating with AI, they unlock its full potential—not just as a time-saver, but as a tool for creating more meaningful and impactful learning experiences.

By navigating the challenges thoughtfully and embracing AI's possibilities, educators can transform their classrooms into dynamic spaces where every student has the opportunity to thrive.

What Are AI Language Models?

The beauty of AI lies in picking the right tool for the job. When you understand what these models can do, it's much easier to match them to your specific needs—whether that's creating lesson plans, translating materials, analyzing data, or digging up credible research. This thoughtful approach doesn't just save you time; it also makes your teaching more impactful and your students' learning experiences richer.

So, what exactly are AI language models? Think of them as highly advanced systems designed to understand and generate human-like text. Models like ChatGPT, Google Gemini, Claude, and Perplexity.ai are powered by machine learning. They analyze massive amounts of language data to produce responses that feel natural and relevant to the context. These tools are all about interpreting the nuances of language—syntax, meaning, and tone—to handle tasks ranging from answering simple questions to generating complex, tailored content.

Each model has its own strengths, making them better suited for certain educational tasks. Let's break down what they're good at and how you can put them to work in your classroom.

The Models and Their Strengths

- **ChatGPT:**
  - *What It Does Best:* ChatGPT specializes in creating natural, conversational responses and breaking down complex topics into simpler terms. It's ideal for brainstorming, drafting lesson plans, and creating student-friendly explanations.
  - *How It Works:* Built on OpenAI's GPT architecture, ChatGPT uses billions of parameters to understand and generate text, making it versatile and highly adaptable to instructional needs.

  - *Best Use Cases:* Designing creative lesson plans, drafting personalized feedback for students, and generating example scenarios for classroom activities.
- **Google Gemini:**
  - *What It Does Best:* Gemini excels in tasks that require summarization, translation, or fact-based accuracy. Its ability to handle multilingual contexts makes it a valuable tool for diverse classrooms.
  - *How It Works:* Using Google's expertise in natural language processing, Gemini integrates advanced capabilities in summarization and multilingual translation with robust search and information retrieval.
  - *Best Use Cases:* Translating lesson materials for English Language Learners (ELLs), summarizing lengthy academic articles, or preparing concise study guides.
- **Claude:**
  - *What It Does Best:* Claude is particularly strong at handling extensive text inputs and structured data. It shines in analytical tasks, such as extracting trends from datasets or summarizing long documents.
  - *How It Works:* Claude leverages advanced language processing to work with "book-size" inputs, focusing on clarity and logical organization in its outputs.
  - *Best Use Cases:* Analyzing large datasets, generating summaries of detailed reports, and structuring student research projects.

- **Perplexity.ai:**
  - *What It Does Best:* Perplexity.ai blends conversational AI with research capabilities, delivering fact-checked responses supported by credible sources. It's highly effective for research-focused tasks.
  - *How It Works:* Powered by GPT-4, Perplexity.ai enhances traditional language modeling with integrated information retrieval and citation tools to verify data accuracy.
  - *Best Use Cases:* Conducting research for classroom discussions, finding peer-reviewed articles, and retrieving reliable data for student projects.

Case Study:

A Climate Change Unit

Determining the best AI language model to use starts with understanding your specific teaching needs. Let's consider how a high school teacher developing a unit on climate change might evaluate and apply different models effectively:

I. **Clarify Your Objective:** Begin by identifying your goals. Are you looking to create engaging lesson plans, provide multilingual support, analyze complex data, or source reliable research? Knowing what you want to achieve helps narrow down which tool is most appropriate.

II. Test for Lesson Planning – ChatGPT:

A. The teacher starts with a broad objective: creating a three-week unit that includes interactive discussions and experiments. Using ChatGPT, they prompt:

"Design a lesson plan for high school students on climate change, incorporating hands-on activities, debates on renewable energy, and an assessment plan."

B. ChatGPT quickly drafts a cohesive outline, breaking down activities by week and providing creative ideas like modeling carbon cycles and simulating climate negotiations. Its strength lies in generating student-centered, creative plans.

III. Evaluate Multilingual Needs – Google Gemini:

A. To ensure accessibility for English Language Learners (ELLs), the teacher uses Google Gemini to translate the lesson materials into Spanish and Mandarin. With a simple command like:

"Translate this activity sheet on greenhouse gases into Spanish for a 10th-grade audience."

B. Gemini delivers accurate translations while simplifying technical terms for clarity. It excels at breaking down complex content for multilingual classrooms.

IV. Explore Data Analysis – Claude:

A. For a project involving CO_2 emission trends, the teacher turns to Claude. By feeding in a dataset and prompting:

"Summarize three key trends in global CO_2 emissions from

1950 to 2020 and highlight significant events affecting these trends."

B. Claude processes the information efficiently, producing a well-organized summary that teachers can incorporate into classroom discussions or presentations.

V. Source Credible Research – Perplexity.ai:

A. When preparing a lecture or student resource on climate change's impact on biodiversity, the teacher uses Perplexity.ai. A query like:

"Find three peer-reviewed articles on how climate change affects Arctic ecosystems, and summarize their key points."

B. Perplexity.ai not only retrieves reliable articles but also provides succinct summaries, saving the teacher hours of research.

How Teachers Can Choose the Best Model

The superpower of AI is in using the right tool for the right task. When you understand what these models can do, it's easy to match them to your needs—whether it's designing lesson plans, translating materials, analyzing data, or gathering reliable research. This thoughtful approach not only saves time but also frees you to focus on connecting with your students and enriching their learning experiences.

AI language models like ChatGPT, Google Gemini, Claude, and Perplexity.ai are highly advanced systems trained to generate human-like text. By recognizing patterns and understanding context, they provide responses that are both coherent and relevant, making them useful for a variety of teaching tasks.

Each model brings something unique to the table. ChatGPT is perfect for brainstorming creative ideas and simplifying complex topics. Google Gemini stands out in summarizing and translating materials for diverse learners. Claude is excellent for analyzing detailed datasets and organizing complex information. Perplexity.ai specializes in fact-based research, offering credible, verified resources for projects.

By understanding these strengths, you can choose the best model for your classroom, turning AI into a supportive and practical partner. Let's take a closer look at how to make these tools work for you:

1. **Define Your Task:** Whether it's generating ideas, translating materials, analyzing data, or conducting research, start with a clear understanding of your objective.
2. **Match the Model to the Task:**
    a. For creative or conversational tasks like lesson planning, ChatGPT excels.
    b. For multilingual support or concise summarization, Google Gemini is your best bet.
    c. For processing large data sets or extensive text, Claude is the strongest choice.
    d. For finding credible, research-based resources, Perplexity.ai stands out.

3. **Experiment with Prompts:** Test the models by crafting detailed, context-rich prompts. The quality of the response often depends on how clearly you communicate your needs.
4. **Evaluate Output Quality:** Assess the responses for relevance, clarity, and accuracy. Choose the model that delivers results aligned with your teaching style and goals.

By taking these steps, teachers can integrate AI into their classrooms in ways that save time, reduce stress, and enhance their students' learning experiences. AI becomes less of a mystery and more of an essential teaching partner, tailored to the unique challenges and opportunities of education.

Try it yourself: Find the Right Model

Pick a Task: Choose something you often do, like lesson planning, translating materials, or finding resources.

Write a Prompt: Be specific and conversational in your request.

Test the Models: Run your prompt through ChatGPT, Google Gemini, Claude, and Perplexity.ai. Compare their responses for clarity, relevance, and usefulness.

Reflect: Which model worked best for your needs? Which one aligns with your teaching style?

This quick exercise will help you discover how AI can simplify your tasks and enhance your teaching. Try it—you might be surprised!

Strengths, Challenges, and Classroom Applications

AI has the potential to revolutionize how teachers approach their work, taking on repetitive tasks and freeing up time for what really matters—teaching and connecting with students. From grading assignments to suggesting fresh lesson ideas, AI can significantly lighten the load. Need to create differentiated practice exercises for a range of abilities? AI can help. Trying to adapt materials to fit each student's learning style? It's got you covered. It's like having a reliable assistant that's available 24/7.

But let's be real—AI isn't a magic wand. There are hurdles to navigate. For starters, the learning curve can feel steep, especially for teachers new to the technology. Training on how to effectively use AI isn't always readily available, and for under-resourced schools, the cost of implementing these tools can be a barrier. And then there are the ethical questions: How do we protect sensitive student data? How do we avoid relying on AI so heavily that it undermines meaningful teacher-student interactions or reinforces existing biases?

When used thoughtfully, though, AI can be a powerful ally in the classroom. Imagine you're teaching verb conjugation in French. AI can generate tailored practice exercises for students, adjusting the difficulty level based on their progress. Working on quadratic equations in math? AI can provide step-by-step walkthroughs to support struggling learners. For students with special needs, adaptive technologies powered by AI make lessons more inclusive, offering personalized support to ensure every student has a path to success.

It's not just about what AI can do for students, though—it's also a game-changer for teachers. Virtual tutors can answer routine student questions, freeing you to focus on creating engaging, dynamic lessons. AI tools can help you draft lesson plans in minutes, recommend creative approaches to your content, or even provide feedback for student writing. It's like having an extra set of hands that never needs a break.

Of course, the key to all this is balance. AI works best when it's a tool to enhance your teaching, not replace your role as an educator. By combining its strengths with your expertise, you can create a classroom environment that's more efficient, inclusive, and engaging. With a bit of thoughtful integration, AI can transform the way you teach—and the way students learn.

Developing Effective Communication with AI

Let's go back to that story about my colleague. What turned everything around for her was learning to "speak AI." The more conversational and specific your prompts are, the better your results will be. For example, instead of saying, "*Create a lesson plan on photosynthesis,*" try: "*Help me design a 10th-grade biology lesson on photosynthesis, including a lab activity and discussion questions.*" That's going to give you much better results, but not even as good as it can get.

It's all about being clear, direct, and a little creative. By mastering this skill, you'll find that AI becomes less of a mystery and more of a reliable partner in your teaching.

Here's the bottom line: This is as bad as AI is ever going to be, and it's only going to become more integrated into education. If we embrace it now, we can lead the way, shaping how it's used to benefit both teachers and students. If we ignore it, we risk falling behind—and that's a position no one wants to be in.

By staying informed and adaptable, we can make the most of what AI has to offer while addressing its challenges thoughtfully. It's not just about keeping up; it's about setting the pace.

So, what's the takeaway? AI isn't here to replace you—it's here to help. By learning to craft better prompts and understanding how these tools work, you can unlock their full potential. Start small, experiment, and don't be afraid to get a little conversational. You might just be surprised by how much AI can do for you and your students.

Chapter 2: Experimentation with Variations

Small changes in prompts can make a big difference in how AI responds, and for teachers, understanding how to refine prompts is transformative. A well-crafted prompt can mean the difference between receiving a generic answer and generating a detailed, tailored resource that meets your exact needs. Whether you're designing lesson plans, crafting engaging activities, or troubleshooting learning gaps, small tweaks in phrasing, specificity, and structure can profoundly impact the quality of AI-generated responses.

For example, think about a common classroom task, like creating a lesson on photosynthesis. A basic prompt like, "*Explain photosynthesis,*" will give you general information, but it might not be suitable for your students' age or knowledge level. However, refining it to something like, "*Explain photosynthesis to 10th-grade biology students, including a step-by-step diagram and a quiz for review,*" makes all the difference. This specific prompt not only directs the AI to tailor the content for high schoolers but also ensures the output is formatted in a way you can immediately use in your classroom.

This process of refinement is what makes working with AI so powerful. It's not just about asking questions—it's about knowing how to ask the right ones. Experimenting with prompts teaches you how subtle changes, like adding a grade level, specifying a format, or focusing on key concepts, can shape the AI's responses. For instance, adding phrases like "*include common student misconceptions*" or "*format this as a slide deck*" gives the AI even clearer instructions, enabling it to provide a more usable product.

Experimenting with prompts helps you understand how small adjustments—such as including key terms, specifying the format, or focusing on student misconceptions—can shape AI responses. Start with a basic question, refine it step by step, and observe how the output changes. By practicing this skill, you can turn AI into a powerful collaborator, enhancing your teaching and elevating your students' learning experience.

Let's Get Specific

Getting specific with your AI prompts is like seasoning a recipe—it transforms a generic response into something that perfectly suits your needs. Just as the right blend of spices enhances a dish's flavor, the right details in your prompts ensure the AI delivers precisely what you're looking for. A vague or broad prompt might give you a general response, but by adding clarity, focus, and intent, you can guide the AI to generate detailed, actionable, and highly relevant resources for your classroom.

Specific prompts are the key to unlocking the full potential of AI. Think of it as giving the AI a roadmap: the more defined the path, the better the destination. When you include details like your students' grade level, the learning objectives, or even the desired format for the output, you help the AI understand not just what you need but how to present it in a way that's immediately useful. For example, instead of asking for general information, specify your exact goals, such as creating a lesson plan, generating interactive activities, or preparing assessments.

This practice isn't just about receiving better answers—it's about saving time and elevating the quality of your teaching tools. A well-crafted prompt can turn a simple AI response into a comprehensive teaching aid, enabling you to focus more on engaging with your students and less on preparing materials from scratch. With a bit of experimentation and refinement, you'll see how tailoring your prompts creates a partnership with AI that's as effective as it is efficient. Let's say you're a history teacher wanting to create a lesson on Ancient Egypt. A basic prompt might be:

> Give me information on Ancient Egypt.

That's too broad—like walking into an ice cream shop and asking for a scoop. You'll get something, but it might not be what you want. A vague prompt will yield general, possibly unhelpful information. A little detail can go a long way, so let's refine it:

> Create a detailed lesson plan for 7th graders on Ancient Egypt, focusing on pyramid construction. Include key historical figures, building techniques, and cultural significance. Add a short quiz and an interactive activity idea.

See the difference? This version specifies the grade level, topic focus, and desired outcomes. Instead of generic information, you get a lesson plan that's ready to use, complete with engaging activities and assessment tools. The AI now has clear instructions on what to include and how to structure its response. You're guiding the AI to cook up exactly what you need.

Math teachers often need to create resources that engage students and address common misconceptions. Crafting the right prompt can make a huge difference in turning AI into a valuable classroom assistant. For instance, if you're designing a lesson or worksheet to help students master linear equations, you might start with a straightforward question like:

> Solve 2x + 5 = 15

While the AI will provide the solution, it's not particularly useful since, for a teacher, this problem is not that taxing. You're getting a simple answer with no added value. Instead, try refining your prompt to give the AI more direction and purpose:

> Generate a worksheet for solving linear equations like 2x + 5 = 15, with step-by-step solutions for each problem. Include tips for avoiding common mistakes and explanations of key concepts.

This improved prompt turns a single problem into a complete teaching resource. By requesting step-by-step solutions and strategies to address common errors, you gain tools to help students learn effectively and avoid misunderstandings. It's not just about getting the correct answer; it's about creating a resource you can use directly in your lesson or share with your students to tackle potential challenges proactively.

Turning a weak prompt into something effective comes down to three simple factors: being specific, knowing your audience, and deciding on the output format. Start by getting clear on what you're teaching—whether it's pyramid construction in history or solving linear equations in math. Then, think about who you're teaching. Are they 7th graders just starting out, or high schoolers who need more depth? Finally, let the AI know what you want. Do you need a lesson plan, a worksheet, a quiz, or maybe an activity? When you include these details, the AI knows exactly what you're looking for and gives you responses that save time and fit your teaching goals perfectly.

These small tweaks can turn a generic response into a tailored teaching resource. It's about being clear, specific, and aligning the prompt with your teaching goals. Keep experimenting with the phrasing, and watch how the AI serves up exactly what you need!

The Nuts and Bolts of Prompt Engineering

Prompt engineering is both an art and a science. It's about crafting clear, well-structured instructions that guide the AI toward producing responses that are not just accurate but also practical and aligned with your goals. Effective prompts are built around three key elements: **context**, **intent**, and **outcome**.

Context: Set the stage for the AI by defining the subject, student level, or background. This helps the AI tailor its response to your specific needs.

Intent: Clearly state what you want the AI to do—whether it's to explain, summarize, create, or analyze.

Outcome: Specify the desired format or product, such as a worksheet, slide deck, quiz, or activity plan.

Let's look at an example from science. A weak prompt like "*Teach about the solar system*" is too vague, leading to a response that may provide general information but lacks focus or actionable detail. In contrast, a more refined prompt such as "*Create a guide for 8th graders on the solar system, focusing on the planets' sizes and distances from the sun. Include an engaging activity and a worksheet for review*" gives the AI clear instructions about the target audience, specific content, and desired output. This improved prompt ensures the response is not only relevant but also practical and ready for immediate use in the classroom.

The process of refining prompts allows you to explore how small variations, such as changing the tone or level of detail, can significantly impact the AI's output. For instance, shifting from a casual to a formal tone, or from a broad to a specific request, helps you understand how AI interprets and adapts to different styles and complexities.

By experimenting with context, intent, and outcome, you're essentially collaborating with the AI. This not only improves the quality of the responses but also helps you develop a deeper understanding of how to effectively communicate your needs. A well-crafted prompt transforms the AI from a basic tool into a powerful teaching assistant, capable of creating targeted and engaging materials for your classroom.

Let's Fix Some Weak Prompts

When we delve into the realm of prompt engineering, we encounter the challenge of refining and strengthening weak prompts to extract the best possible responses from AI language models. Consider a prompt like:

> Tell me about the French Revolution.

While this prompt is straightforward, it may not provide enough specific guidance for the AI to generate a comprehensive and detailed response.

Now, transform the prompt by being more specific about what were looking for, such as:

> Explain the factors that led to the French Revolution and its impact on France's political landscape.

We introduce a more nuanced and detailed instruction for the AI to follow. This adjustment sets a different expectation for the depth and scope of the response, prompting the AI to delve into specific historical, social, and political aspects related to the French Revolution.

By fixing weak prompts in this manner, we can observe how the AI handles varying levels of complexity and specificity in the instructions it receives. This process not only enhances the quality of the generated content but also provides valuable insights into the AI's ability to interpret and respond to well-crafted prompts.

We can even take this the next step farther. By adding grade levels and how we want the information presented to us, we can better tailor the AI's response to fit our specific needs in that moment:

> Explain the factors that led to the French Revolution and its impact on France's political landscape in a way that would help freshmen in high school better understanding the political landscape. Present the information in the form of a chart for me to use on a side presentation.

Now that we've added a grade level, how we want the information presented, and what we're going to use it for, the AI will produce a more relevant response formatted in a way that will better suit our current needs.

You could keep going here too, getting more specific with each request, seeing how the AI's responses change.

Try it yourself: Refining Your Prompts

Try starting out with a basic prompt and see what the AI give back to you. Then recreate the prompt with increasing levels of specificity and see what you get.

The Magic of AI

Prompt engineering is like setting the stage for a performance—when you craft a well-structured and clear prompt, you're giving the AI the cues it needs to deliver exactly what you're looking for. It's a collaborative process: you nudge the AI in the right direction with context, intent, and expected outcomes, and it responds with tailored, insightful results. Mastering the art of prompt engineering empowers educators to communicate effectively with AI, transforming it from a basic tool into a reliable teaching assistant.

> Mastering the art of prompt engineering empowers educators.

In education, crafting effective prompts isn't just about getting information; it's about encouraging deeper engagement, critical thinking, and active learning. Weak prompts often yield straightforward but shallow responses, missing opportunities for depth or meaningful application. By refining these prompts with clear context, specific intent, and actionable outcomes, we can unlock AI's full potential and elevate it into a valuable classroom partner.

Let's explore how this works with some subject-specific examples, showing how weak prompts can be transformed into powerful tools for teaching. These examples illustrate how a little extra detail can foster richer, more useful AI responses that align with your educational goals.

Science

When crafting prompts for AI, the goal is to go beyond simple requests and create instructions that yield meaningful, actionable results. A weak prompt often lacks the necessary detail to guide the AI effectively, resulting in a generic response that may not align with your needs. Let's explore this with an example:

> List the planets in the solar system.

This prompt will generate a correct list of planets, but it doesn't provide any depth, context, or opportunities for engagement. Without additional detail, the AI can't anticipate the needs of your audience or the purpose of the information:

> Create a detailed guide on the solar system planets, focusing on their unique features and the history of how each was discovered. Include engaging activities for eighth-grade students to explore these differences in a worksheet.

This updated version addresses key elements that the weak prompt misses. It defines the context (eighth-grade students and the solar system), specifies the intent (a detailed guide that includes discovery history and unique features), and outlines a clear outcome (a worksheet with engaging activities). The refined prompt turns a basic query into a directive for creating a rich, interactive learning experience tailored to a specific audience.

English Literature

When crafting prompts, specificity is crucial to ensure the AI generates a response that aligns with your teaching objectives. A weak prompt often lacks the necessary detail to provide anything beyond a generic result, leaving you with something that might require significant revision to be useful. For example, take the following prompt:

> Give me a lesson plan for Shakespeare's "Hamlet."

This prompt is too broad and unfocused. While the AI might provide a general overview or basic plan, the result is likely to be a cookie-cutter response. Now compare that to a refined version:

> Develop a lesson plan exploring the themes of betrayal and revenge in Shakespeare's "Hamlet" intended for honors-level high school seniors, including interactive group discussions and creative writing tasks where students compose a modern-day soliloquy.

This improved prompt eliminates ambiguity by clearly defining the **context**, **intent**, and **outcome**—key components of an effective prompt. By focusing on betrayal and revenge, it narrows the scope to align with specific themes in *Hamlet*, and specifying "honors-level seniors" ensures the material is appropriately challenging.

The **intent** outlines the desired learning activities, such as group discussions to encourage collaborative analysis and a creative writing task, like composing a modern soliloquy. These elements promote deeper understanding and critical thinking while connecting the material to students' perspectives.

Finally, the **outcome** specifies a well-rounded lesson plan with analytical and creative components, transforming the AI's role from merely providing information to generating a comprehensive teaching resource. By addressing all three components, the refined prompt ensures the AI produces tailored, actionable results that engage students and align with classroom goals. This strategy empowers educators to enhance learning across disciplines by using AI as an effective teaching partner.

Social Studies

When crafting prompts for AI, it's important to move beyond vague instructions to provide the clarity needed for meaningful, tailored responses. Consider a social studies teacher developing a world history lesson for high school freshmen. They might start with a basic prompt like:

> Produce a lesson on the causes of World War I.

While this prompt will generate a response, it's too broad and lacks the detail required to create a focused, engaging lesson. The AI might provide a generic overview of the causes, but it won't address specific teaching needs or offer interactive resources. A prompt like this is functional but misses the opportunity to leverage AI's full potential.

To improve, we need to incorporate **context**, **intent**, and **outcome**. Context focuses the content, intent specifies the purpose or approach, and outcome defines the desired format or final product. For example, the revised prompt could be:

> Produce a series of lesson slides that dissect the complex web of alliances and political tensions leading to World War I, complemented by primary source analysis activities for students.

This improved version provides much clearer instructions. The **context** highlights the causes of World War I and introduces a specific focus on alliances and political tensions. The **intent** specifies the need for primary source analysis to foster critical thinking. Finally, the **outcome** requests the creation of a slide deck, offering a ready-to-use resource for classroom teaching.

This refined prompt doesn't just ask for information—it guides the AI to create an engaging, comprehensive educational experience that aligns with the teacher's goals. By being specific and intentional, the AI becomes a valuable partner in developing rich, interactive content that goes beyond a basic lesson.

Mathematics

When crafting prompts, the goal isn't just to get the correct answer (unless it is)—more often it's to foster deeper understanding, critical thinking, and meaningful engagement. A poorly designed prompt might deliver the right response but often overlooks the opportunity to clarify concepts, address common misconceptions, or connect ideas to broader applications. For example, let's consider this straightforward request:

> Solve the equation sin(x) = 0.5.

While the AI will provide the solution, such as x = arcsin(0.5) and any relevant periodic solutions, this prompt does little to enhance understanding. It doesn't explain how the solution relates to the unit circle, trigonometric properties, or the broader context of trigonometric equations. Now compare this to a more detailed and purposeful prompt:

> Explain the process of solving the equation x + 5 = 10, and discuss how this method can be applied to similar algebraic problems. Additionally, present the three most common errors students make while learning this process and how I could teach them to avoid those issues.

This improved prompt delivers much more. The **context** specifies solving trigonometric equations within a specific interval, ensuring the solution is tailored to a classroom setting. The **intent** extends beyond simply solving the equation to explaining the reasoning behind the solution, identifying common pitfalls, and addressing instructional strategies. The **outcome** includes not only a clear explanation but also a related problem to reinforce the concept, providing a resource that is both educational and practical.

This refined prompt transforms the AI into a teaching assistant capable of offering detailed explanations, preemptively addressing student challenges, and generating supplemental materials for deeper engagement. It goes beyond a simple answer, equipping the teacher with tools to enhance student understanding and avoid common mistakes, such as forgetting additional solutions or misapplying the periodicity of trigonometric functions.

By crafting prompts with context, intent, and outcomes, educators can elevate the role of AI from a basic problem solver to a collaborative partner in teaching. For example, specifying the level of the students—like high school juniors in a trigonometry class—or requesting a visual representation, such as a unit circle diagram, further refines the AI's output to align closely with the teacher's needs. This approach encourages critical thinking, builds conceptual understanding, and creates a richer learning experience for students.

Now it's your turn! Try applying these principles to your own teaching scenarios and see how refining your prompts can elevate your AI interactions.

Case Study:

Refining AI Prompts for a Science Lesson

I. **Background:** Ms. Carter, a middle school science teacher, wants to use AI to help create a lesson plan on ecosystems. Her goal is to introduce her 7th-grade students to the concept of food chains, emphasizing key organisms' roles and their interdependence within an ecosystem. Initially, she tries a basic prompt but finds the response too general to meet her needs.

II. **Initial Prompt:** *Create a lesson plan about ecosystems."*

III. **Result:** The AI generates a basic lesson that briefly describes ecosystems, listing producers, consumers, and decomposers with minimal detail. While the response is factually correct, it lacks structure, engaging activities, and depth suitable for her students.

IV. **Refined Prompt:** *Design a 45-minute lesson plan for 7th graders on ecosystems, focusing on food chains and the interdependence of organisms. Include examples from a forest ecosystem, an interactive group activity to create their own food chains, and a 5-question quiz to assess understanding."*

V. **Result:** The revised prompt yields a detailed response that includes a structured lesson. The AI provides:

A. A brief introductory explanation of food chains.

B. Examples of forest ecosystems, such as interactions between wolves, deer, and plants.

C. Instructions for a hands-on activity where students create and present their own food chains using provided cards with organisms.

D. A five-question quiz to evaluate key concepts like producers, consumers, and decomposers.

VI. **Why It Worked:** The refined prompt applies the **Context, Intent, Outcome** framework:

A. **Context:** Specifies the focus (food chains in ecosystems) and the audience (7th-grade students).

B. **Intent:** Clearly requests a lesson plan with an interactive activity and assessment.

C. **Outcome:** Defines the format (structured lesson plan with examples, an activity, and a quiz).

VII. **Reflection:** By experimenting with her prompts, Ms. Carter discovered how detailed instructions can guide the AI to create tailored, practical resources. She saved significant time planning and developed a more engaging and effective lesson for her students.

VIII. **Takeaway for Educators:** When using AI, a clear and specific prompt can transform a simple request into a robust teaching tool. Try refining your prompts by focusing on Context, Intent, and Outcome to get responses that align perfectly with your classroom goals.

How the Formula Talks to the AI

The *Experimentation with Variations* formula serves as a structured framework for communicating with AI language models. It involves providing a well-crafted prompt to the AI and observing the resulting changes in the generated responses. This process guides the AI in understanding the nuances of prompts and their impact on the nature and quality of the generated content.

When utilizing this formula, educators, AI developers, and researchers gain practical experience in comprehending how subtle alterations in prompts can influence AI responses. By inputting carefully constructed prompts and observing the corresponding variations, they enhance their ability to craft effective and impactful prompts for diverse educational and research contexts.

Prompt engineering, as a multi-level approach to problem-solving with AI, is not just about formulating questions; it's about encoding context, intent, and expected outcomes into concise and effective prompts. This approach serves as a powerful tool to simplify or expand the task at hand, offering a flexible and enjoyable approach to problem-solving.

As AI technologies become more integrated into various domains, prompt engineering plays a crucial role in maximizing model performance, ensuring safety, conserving resources, and improving domain-specific outputs. It is essential for the effective utilization of large language models in natural language processing and will continue to grow in importance as AI becomes increasingly integrated into daily life.

Try it yourself: Refining Your Prompts II

Use **Context, Intent, Outcome** to develop your own prompts.

- Set the **context** by considering a topic your students are struggling with, their age, and their level.
- **Intent** is crucial decide what the chatbot should help them understand or explore.
- Determine the **outcome**, what product should the chatbot create that would be most helpful to you.

Run the prompt you developed through a chatbot to see what you get. Try a simplified version of the prompt to compare the results.

Chapter 3: Balancing Specificity and Clarity

Creating effective AI prompts is both an art and a science, requiring a careful balance between specificity and clarity. A well-crafted prompt serves as a guide for the AI, ensuring its responses align closely with your educational goals. This chapter builds on the foundations laid earlier, focusing on how educators can refine their prompts to achieve meaningful, targeted responses that enhance teaching and learning.

Diving into the process, we'll explore how adjustments in detail and focus influence AI outputs, demonstrating how educators can unlock the full potential of AI tools. Starting with broad questions, we'll gradually introduce more specific and clear elements, observing the impact of these refinements on the AI's responses. Through hands-on activities and real-world examples, this chapter aims to empower educators with the skills needed to craft precise, actionable prompts that yield engaging, accurate, and practical learning resources.

The goal is to move beyond generic outputs and ensure that every interaction with AI delivers value. By mastering specificity and clarity, teachers can fine-tune prompts to align with their curriculum needs, address student challenges, and foster deeper understanding. Whether it's creating lesson plans, designing assessments, or generating interactive activities, thoughtful prompt engineering transforms AI into a powerful ally in the classroom.

Through a series of practical exercises and examples, educators will learn to create prompts that balance being neither too vague nor overly restrictive. By the end of this chapter, you'll understand the critical role that clarity and detail play in leveraging AI effectively, ensuring prompts lead to responses that are not only informative but also aligned with your teaching objectives.

Crafting Precise Prompts

Creating effective AI prompts is both an art and a science, requiring a balance between specificity and clarity. These elements directly influence the quality of AI-generated responses, making them essential for educators maximizing AI tools in their classrooms. A well-crafted prompt incorporates context (subject matter and audience), intent (task purpose), and outcome (desired product, like a quiz or lesson plan). This structure transforms generic outputs into actionable teaching resources.

Consider the difference between vague and refined prompts. A weak prompt like "*Explain photosynthesis*" may yield a general response but lacks focus and relevance. A more detailed prompt such as "*Explain the process of photosynthesis to 10th-grade biology students, emphasizing the role of chlorophyll and light energy in converting carbon dioxide and water into glucose and oxygen. Include a simple diagram and a follow-up quiz with five questions,*" provides clear instructions. This refined prompt integrates context, intent, and outcome, ensuring targeted, accurate, and classroom-ready responses.

Specificity and clarity apply across subjects. For instance, in history, "*Discuss the causes of World War I*" might produce a surface-level response. Refining it to "*Create a lesson plan for 10th graders explaining the causes of World War I, focusing on alliances, nationalism, and militarism. Include an interactive timeline activity and discussion questions,*" ensures the response directly supports teaching goals and fosters engagement.

Crafting prompts goes beyond asking questions; it's about incorporating precise details into concise instructions. For example, instead of "*Explain climate change,*" try "*Compose a 700-word article for high school seniors explaining the causes and effects of climate change. Focus on recent data, include real-world examples, and provide a call to action for sustainable practices.*" Specificity ensures relevant, actionable content.

Crafting effective prompts takes practice, but the rewards are well worth it. By refining prompts with clear context, intent, and outcomes, you can turn AI into a valuable partner for creating engaging and practical teaching resources. Now, it's your turn to experiment and see how small changes can make a big difference in AI responses.

Try it yourself: Precision Prompts

Start with a general prompt. Refine it incrementally by adding details about the grade level, subject specific terms, or practical teaching goals. Observe how the AI's response improves.

Balancing Specificity and Flexibility

Creating effective AI prompts requires a delicate balance, much like tuning an instrument to produce the perfect melody. Too general, and the response risks becoming a shallow overview that lacks focus or depth, leaving educators with little actionable value. On the other hand, prompts that are overly specific can suffocate the AI's flexibility, producing rigid responses that may miss broader connections or fail to inspire creative engagement. This balance is especially crucial in educational settings, where the goal is not just to retrieve information but to foster critical thinking, creativity, and deeper understanding.

By striking the right balance between specificity and flexibility, educators can guide AI to generate thoughtful, actionable, and comprehensive outputs tailored to their needs. A balanced prompt acts as a roadmap, offering enough direction to keep responses on track while leaving room for nuanced exploration. For instance, a well-crafted prompt can ensure that AI delves into a topic deeply while also presenting broader implications, making it an invaluable tool for creating lesson plans, classroom activities, or teaching aids.

This delicate equilibrium empowers educators to use AI as a partner in enhancing learning experiences. Instead of settling for generic or overly constrained responses, teachers can craft prompts that align with their educational objectives, spark curiosity, and engage students in meaningful ways. The process of achieving this balance is not just about efficiency—it's about unlocking the potential of AI to enrich teaching and learning, making it more adaptive, innovative, and impactful.

The Risks of Vague Prompts

When a prompt is overly broad, the AI is left to interpret the user's intent without clear guidance. This often results in responses that are generic, unfocused, and insufficiently tailored to meet specific educational goals. For example, consider the prompt:

> Discuss the impact of technology.

This lacks essential details, offering no direction on what aspect of technology should be discussed. Should the response focus on historical advancements, societal shifts, economic transformations, or something else entirely? Without a defined scope, the AI might provide a surface-level overview that touches on multiple topics but fails to explore any of them in meaningful depth. Such a response might skim over key points without aligning closely with the classroom's learning objectives, leaving educators with content that requires significant refinement before it can be used effectively.

In educational settings, vague prompts squander the potential of AI as a tool to deliver tailored and actionable insights. Students, for instance, might be left with broad information that neither challenges their critical thinking nor aligns with their curriculum. Teachers looking to save time or gain new teaching perspectives may instead find themselves with a response that requires additional effort to clarify or reframe. By failing to provide the AI with enough direction, educators miss an opportunity to generate materials that are not only accurate but also deeply relevant to their instructional needs.

Crafting prompts that eliminate vagueness is essential to unlocking the AI's potential for creating engaging, specific, and classroom-ready resources. Clear guidance transforms the AI from a general information provider into a focused assistant capable of supporting precise learning outcomes.

The Pitfalls of Overly Specific Prompts

While specificity is crucial in crafting effective AI prompts, an overly detailed prompt can backfire by narrowing the scope so much that it stifles the AI's ability to generate comprehensive insights. By focusing too narrowly, educators risk receiving responses that lack the broader context needed to engage students or meet diverse learning objectives.

Overly specific prompts limit the AI's ability to provide well-rounded responses, narrowing its focus. Instead of fostering connections or offering diverse perspectives, overly rigid prompts limit the AI, reducing its usefulness as a teaching tool, for example:

> List five economic policies enacted in France during World War I and explain how each affected rural wheat farmers.

This prompt is highly focused and will likely generate precise information about the economic policies and their impact on a very specific demographic. However, such precision comes at the expense of broader educational value. The AI might ignore other vital aspects of the war's economic landscape, such as its effects on urban populations, industrial workers, or international trade relations. Without this context, the response might fail to provide students with a holistic understanding of World War I's economic repercussions, making it less effective as a teaching resource.

Additionally, the rigid structure of the prompt can limit the AI's capacity for creative exploration or connections between related topics. For instance, examining the broader interplay between economic policies and societal shifts during the war could help students develop a deeper, more nuanced understanding of the period. By boxing the AI into a narrowly defined task, educators may inadvertently miss opportunities for richer, multidimensional insights that could elevate their lesson plans.

Overly specific prompts also pose challenges when the teaching objective involves encouraging critical thinking or drawing comparisons. For example, asking students to analyze how World War I's economic impacts varied across different regions or demographics would require a broader scope than what the narrow prompt allows. The lack of flexibility in overly detailed prompts can hinder the AI's ability to serve as a versatile teaching assistant capable of generating adaptable and engaging content.

Striking the right balance between specificity and flexibility ensures that prompts provide enough direction to generate relevant responses while allowing room for comprehensive, contextual, and creative outputs. This balance is key to leveraging AI effectively in educational settings.

Finding the Balance

Striking the right balance in a prompt is key to guiding the AI effectively while leaving room for creative and meaningful exploration. A well-balanced prompt provides clear direction without overloading the AI with unnecessary constraints, enabling it to generate responses that are both detailed and adaptable. This approach ensures that the results are relevant, comprehensive, and directly aligned with the educator's teaching goals.

For example, consider the following balanced prompt:

> Explain the social and economic impacts of World War I in Europe, focusing on rural communities and industrial workers. Include examples from two countries and suggest a classroom activity for analyzing these impacts.

This version offers specific guidance by defining the context (social and economic impacts of World War I), the audience (students analyzing rural and industrial dynamics), and the desired outcomes (historical comparisons and actionable classroom activities). Yet, it avoids excessive rigidity, allowing the AI to explore a variety of angles, such as comparing the effects in France and Germany or proposing different interactive learning strategies.

By maintaining this balance, educators can harness AI to produce responses that are not only comprehensive but also flexible enough to adapt to diverse classroom scenarios. The result is a teaching resource that engages students with thoughtful content and interactive opportunities for deeper learning.

A Guide to Precision and Flexibility

When working with AI to generate educational resources, finding the balance between specificity and flexibility is a vital skill. Too vague, and the prompt risks producing generalized, unhelpful responses. Too detailed, and it can stifle the AI's ability to generate diverse and creative outputs. The sweet spot lies in crafting prompts that guide the AI clearly while leaving room for it to explore different dimensions of the topic. This balance enables educators to receive responses that are both insightful and adaptable to a variety of teaching needs, offering tools that are not only relevant but often exceed initial expectations.

To achieve this, consider how prompts function as a conversation starter. Just as an open-ended yet focused question invites a more engaging dialogue, a well-crafted prompt encourages the AI to delve deeply into a topic while staying relevant to the task at hand. By balancing structure with creativity, educators can tap into the AI's potential to suggest fresh approaches or unexpected solutions to common challenges. This balance ensures that the outputs align with your educational goals while allowing the AI to bring new perspectives or ideas you might not have initially considered. The following tips can help educators refine their prompts for optimal results and make the most of AI's capabilities in the classroom.

Tips for Achieving the Right Balance

Crafting balanced AI prompts ensures relevant and adaptable responses. Start with a broad topic, refine with details to focus on your goals, and leave room for creativity. The tips below will help you achieve this balance effectively.

5. **Start with Broad Prompts:** Begin by framing the general topic to establish the scope. For example: _"Discuss key labor laws in 19th-century England."_
6. **Add Layers of Specificity:** Gradually refine the prompt to target specific aspects while maintaining flexibility. Enhance it to: _"Summarize key labor laws in 19th-century England that shaped factory working conditions and highlight their effects on worker rights. Include a classroom activity analyzing historical perspectives."_
7. **Leave Room for Exploration:** Avoid boxing the AI into overly rigid parameters. Instead of demanding exhaustive detail on one narrow aspect, provide prompts that encourage broader synthesis.

By following these steps, you can create prompts that provide the AI with enough structure to generate useful responses while retaining the flexibility needed for innovative and comprehensive results.

Applying Balanced Prompts in Practice

After refining your prompts, take a moment to assess the AI's responses. Does the output align with your teaching objectives? Are there any surprising insights or creative elements that enhance its value as a classroom resource? By consistently reviewing and adjusting your prompts, you can fine-tune your approach and gain deeper insights into how AI interprets and responds to various levels of instruction.

For instance, imagine crafting a history lesson prompt and receiving not only a detailed timeline of events but also suggestions for interactive student activities, such as debates or role-playing scenarios. This type of response demonstrates how a well-balanced prompt can inspire engaging and diverse teaching strategies. With practice, you'll find that mastering the balance of specificity and flexibility allows you to unlock the full potential of AI as a powerful tool for education.

By mastering the art of crafting balanced prompts, educators can ensure their AI tools provide responses that are insightful, flexible, and tailored to their unique classroom needs. Keep experimenting with specificity and flexibility, and let the AI become a powerful ally in shaping innovative and engaging learning experiences.

Case Study:

Incremental Refinement in Prompt Crafting

I. **Scenario:** Ms. Lopez, a high school math teacher, is preparing resources for her trigonometry unit. She decides to use AI to generate teaching materials but struggles to achieve the level of detail and relevance her students need.

II. **Initial Prompt:** "*Solve a trigonometry problem.*"

   A. *Basic Example:* The AI solves the equation, like sin(x) = 0.

   B. *Issue:* The response provides a direct solution without context or explanation.

   C. *Lack of Depth:* Minimal educational value as it skips over the reasoning or steps involved.

   D. *Missed Opportunity:* Does not include teaching aids, common mistakes, or broader applications.

III. **Refined Prompt:** "*Solve a trigonometry problem involving sine and cosine.*"

   A. *Improved Prompt:* The AI now solves equations involving both sine and cosine.

   B. *Increased Relevance:* The response focuses on both trigonometric functions, providing a more targeted answer.

   C. *Issue:* The output remains generic, lacking specific strategies or detailed explanations.

   D. *Missed Opportunity:* Does not address common student errors or provide instructional guidance for educators.

IV. **Targeted Prompt:** "*Create a step-by-step guide for solving trigonometric equations like sin(x) = 0.5 and cos(x) = -0.5 over the interval \[0, 2π\]. Include common mistakes students might make and provide strategies to avoid them.*"

   A. *Comprehensive Output:* The AI provides a detailed step-by-step solution process.

   B. *Error Identification:* Highlights common mistakes, such as overlooking all solutions within the given interval.

   C. *Instructional Support:* Suggests practical strategies to help students avoid and correct these errors.

   D. *Educational Value:* Creates a robust teaching tool that enhances both understanding and application for students.

V. **Outcome:** By iteratively refining her prompts, Ms. Lopez transforms the AI into a valuable teaching assistant, producing targeted, detailed, and actionable resources that directly address her classroom needs.

VI. **Key Takeaway:** Start with a general prompt and refine it incrementally. Each layer of specificity adds clarity and context, ensuring the AI delivers tailored responses that align with your educational goals.

Applying Incremental Refinement

Building better prompts is an iterative process. By starting broad and gradually adding layers of detail, educators can guide AI to deliver responses that align with specific learning objectives. Each refinement ensures greater clarity and relevance, transforming basic outputs into actionable classroom tools.

Try it yourself: Incremental Refinement

Start with a broad question from your subject area and refine it step by step. Observe how the AI's responses evolve and improve with each iteration.

Reverse Engineering Prompts

Another way to hone your prompt-writing skills is through reverse engineering. Imagine you receive an AI-generated response that provides a detailed breakdown of photosynthesis, complete with a labeled diagram and a quiz. By analyzing the output, you might deduce the original prompt was something like:

> **Explain photosynthesis to high school biology students, including a labeled diagram and a 5-question quiz focused on the role of chlorophyll.**

Reverse engineering helps educators understand how specific and clear prompts shape AI outputs. By reconstructing and refining these prompts, you can discover how to craft more effective instructions to achieve your goals.

Try it yourself: Reverse Engineering

Take an AI-generated response and reconstruct the prompt that might have produced it. Refine the prompt further to see if you can improve the output.

Contextual Adjustments

Contextual adjustments are another powerful strategy for refining prompts. For instance, a history teacher initially asks, "*Describe a historical event.*" The AI delivers a generic response with little depth. Adjusting the context to, "*Describe the events leading to the signing of the Treaty of Versailles in 1919, focusing on the perspectives of Germany and France,*" transforms the response into a focused, nuanced exploration of a pivotal moment in history. This approach ensures the AI tailors its output to meet specific educational needs, engaging students with precise and relevant content.

Try it yourself: Contextual Adjustments

Take a general prompt and modify the context—such as changing the time period, subject focus, or audience level. Observe how the AI adapts its responses and note the improvements in relevance and depth.

Enhancing Engagement Through Tailored Prompts

Combining specificity, clarity, and context allows educators to craft prompts that turn AI into a dynamic tool for creating engaging learning experiences. For example, a teacher looking to cover Shakespeare might start with:

> Create a lesson on Shakespeare.

While functional, this lacks depth. Refining the prompt to:

> Develop a lesson plan for high school seniors exploring the themes of ambition and guilt in Shakespeare's Macbeth. Include discussion questions, a creative writing activity, and a quiz on key scenes.

yields a comprehensive, ready-to-use resource tailored to classroom objectives.

By iterating on prompts, leveraging reverse engineering, and adjusting context, educators can maximize the value of AI tools, ensuring they generate high-quality, actionable teaching resources that align with specific goals and enhance student engagement.

Mastering the art of prompt crafting allows educators to fully harness the capabilities of AI. Through clarity, specificity, and a willingness to iterate, teachers can create tailored tools that enhance learning and save valuable time. Dive into the exercises in this chapter to refine your skills, experiment with new ideas, and elevate your AI-assisted teaching to the next level.

Try it yourself: Peer Review

Write a series of prompts and exchange them with a colleague. Provide feedback on clarity, specificity, and adaptability, then revise based on their insights.

Chapter 4: Incorporating Context Appropriately

Context is the cornerstone of effective prompt engineering, shaping the relevance, accuracy, and utility of AI-generated responses. By embedding clear and detailed context into prompts, educators create a framework that clarifies intent, aligns tasks with specific objectives, and minimizes ambiguity. This guidance ensures the AI interprets the request within the appropriate academic domain, grade level, or real-world scenario, making outputs actionable teaching resources rather than generic responses.

When crafting prompts, context serves as the key to unlocking precise and meaningful AI responses. A clear framework not only enhances the relevance of the output but also ensures alignment with specific educational goals. Consider a history teacher exploring the Renaissance. Beginning with a vague prompt like:

> Summarize events from the Renaissance.

Such a request could yield a broad response, touching on diverse topics without depth or focus. This lack of direction leaves the AI guessing at what is most relevant. However, refine the prompt to:

> Summarize key events of the Italian Renaissance, focusing on the Medici family and their arts patronage.

the teacher embeds critical context. This revision narrows the scope and specifies the subject, guiding the AI to deliver a response that is more tailored, insightful, and classroom-ready. By clearly defining the topic, focus, and desired angle, the improved prompt illustrates how adding detail transforms a general question into an effective teaching tool.

When it comes to crafting effective AI prompts, context does more than shape relevance—it safeguards accuracy and mitigates potential pitfalls such as misinformation or bias. By clearly defining parameters like time periods, geographic regions, or focal concepts, educators guide the AI to deliver precise and purposeful responses. For example, a prompt such as:

> Explore the impacts of the Harlem Renaissance on African American literature and culture, highlighting key figures like Langston Hughes and Zora Neale Hurston.

provides a clear framework. This ensures the response focuses on the intended topic without straying into unrelated Renaissance periods or offering irrelevant details. The explicit context transforms the prompt from a broad question into a targeted tool, reducing ambiguity and enhancing the quality of the AI's output.

Context plays a pivotal role in guiding AI to generate reliable and focused responses, particularly by clarifying the boundaries of its knowledge. By defining what information is relevant or expected, educators can prevent inaccuracies and irrelevant content, ensuring outputs remain precise and trustworthy. For example, a science teacher might refine a prompt to:

> Explain the process of photosynthesis in desert plants, focusing on adaptations like CAM photosynthesis.

This detailed prompt directs the AI to address a specific aspect of photosynthesis, tailoring the response to align with the educator's goals while avoiding unrelated or overly general explanations.

Embedding context in prompts not only enhances the accuracy and relevance of AI responses but also promotes inclusivity and fairness. By aligning prompts with specific educational objectives and diverse perspectives, educators can guide AI to deliver more balanced and meaningful content. For instance, a prompt such as:

> Develop a lesson plan for 10th-grade students on the economic and social impacts of World War II, incorporating perspectives from multiple countries.

ensures the AI explores the topic in a comprehensive and unbiased manner, fostering a richer, more inclusive learning experience.

Embedding context in prompts is essential for creating AI responses that are accurate, relevant, and inclusive, aligning with specific educational goals while fostering comprehensive learning experiences. Context provides the AI with a clear framework to interpret tasks, reducing ambiguity and ensuring outputs are tailored to the desired subject, grade level, or objective. By clarifying the scope and intent of a prompt, educators can mitigate common pitfalls such as generic responses, misinformation, or bias, turning AI into a reliable teaching tool. This approach not only enhances the precision and depth of responses but also encourages diverse perspectives, enriching classroom discussions and promoting critical thinking. When educators thoughtfully integrate context into prompts, they ensure the AI's output is not just information-rich but also meaningful and aligned with pedagogical best practices, ultimately transforming how content is delivered and engaged with in the classroom.

Experimenting with Context in Different Subjects

Context plays a pivotal role in shaping the relevance and utility of AI-generated responses across diverse academic disciplines. Whether crafting lesson plans, developing quizzes, or generating engaging classroom discussions, the context embedded in a prompt ensures that the AI aligns its output with specific teaching goals. A history teacher might focus on a particular time period, like the Renaissance, while a science teacher could target molecular processes in photosynthesis. By tailoring prompts to include clear contextual elements—such as grade level, specific topics, or learning objectives—educators can unlock the AI's potential to produce tailored, actionable resources. This alignment between context and intent transforms AI from a general information provider into a tool that creates meaningful, subject-specific content.

The power of context lies in its ability to narrow the scope of AI outputs without stifling creativity, ensuring that responses are both detailed and flexible. By experimenting with context, educators can refine their prompts to meet varying levels of complexity or specificity, depending on their students' needs. For example, shifting the focus of a prompt from general Renaissance history to the influence of the Medici family allows the AI to zero in on a specific aspect, producing insights that enrich the learning experience. The following case study illustrates how context can elevate the quality and relevance of AI-generated materials in a high school history class.

Case Study:

Embedding Context for Effective AI Responses

I. **Scenario:** Ms. Carter, a high school history teacher, wants to use AI to create lesson materials on the Renaissance. Her goal is to generate resources that align with her curriculum and engage her students.

II. **Vague Prompt**

   A. *Prompt: "Summarize events from the Renaissance."*

   B. *Output:* The AI provides a broad overview, touching on unrelated aspects of the Renaissance, such as developments in art, science, and politics across multiple regions and timeframes.

   C. *Analysis:* While the response includes accurate information, it lacks focus and specificity, making it challenging for Ms. Carter to adapt it directly for her classroom. Students could feel overwhelmed by the generality of the content.

III. **Improved Prompt**

A. *Prompt: "Summarize key events of the Italian Renaissance, focusing on the Medici family and their patronage of the arts."*

B. *Output:* The AI generates a concise summary of the Medici family's contributions, including their influence on artists like Michelangelo and Leonardo da Vinci, along with an overview of their political and cultural impact during the Italian Renaissance.

C. *Analysis:* This refined prompt provides clarity and a narrowed scope, ensuring the AI delivers content relevant to the lesson's objectives. The information is directly applicable to Ms. Carter's classroom needs and enhances the depth of her students' understanding.

IV. **Benefits of Embedded Context**

A. *Focus:* Adding context ensures the AI concentrates on specific aspects of the subject, making responses more relevant and actionable.

B. *Clarity:* Clear prompts reduce ambiguity, guiding the AI to generate outputs aligned with educational goals.

C. *Relevance:* Tailoring the prompt to the intended audience or learning level helps create materials that are immediately useful in the classroom.

V. **Summary:** This case study illustrates the transformative power of embedding context in AI prompts. By shifting from a vague to a focused prompt, Ms. Carter ensures the AI's output is precise, relevant, and aligned with her teaching objectives. Incorporating clear context not only saves time but also enhances the learning experience for students, allowing them to engage deeply with the subject matter.

Guidelines for Balancing Context

Balancing specificity and flexibility is essential for crafting effective AI prompts. The level of detail in a prompt should align with the complexity of the subject matter. For straightforward topics, minimal context may suffice, allowing the AI to produce relevant yet adaptable outputs. However, advanced or nuanced topics demand greater specificity to guide the AI toward generating responses that are both accurate and comprehensive. Striking this balance ensures that prompts neither overwhelm the AI with unnecessary detail nor leave it adrift with insufficient direction.

Prompts must also align closely with learning objectives to maximize their relevance and utility. Including critical details such as the grade level, subject focus, and desired outcomes ensures that the AI's responses are tailored to specific teaching goals. This alignment transforms AI outputs into actionable educational tools that directly support lesson plans and enhance classroom engagement.

Considering the AI's existing knowledge base is another crucial factor in prompt crafting. For topics that are well-documented, the AI may require only basic context to generate accurate responses. Conversely, niche or advanced subjects necessitate more detailed prompts to provide the AI with a clear framework for producing informed and relevant content.

Understanding this balance allows educators to adjust their prompts to suit the depth and breadth of information needed.

Effective prompt engineering also benefits from iterative refinement. Starting with a broad prompt and gradually adding layers of detail allows educators to test how changes in context influence the AI's output. This step-by-step process not only improves the clarity and relevance of the prompt but also ensures that the AI's responses evolve to meet specific needs.

All that said, avoiding prompt overload is critical to maintaining flexibility while providing sufficient guidance. Overly detailed prompts can stifle the AI's ability to generate creative and diverse responses, while overly vague ones risk yielding irrelevant results. Clear and balanced prompts offer enough structure to guide the AI without constraining its capacity for generating insightful and comprehensive outputs.

Practical Exercises for Educators

Adding layers of context to a prompt is an essential exercise for understanding how specificity enhances AI-generated responses. This activity allows educators to observe how slight changes to a prompt—such as specifying the audience or focus—can dramatically improve the relevance and depth of the output. By starting with a broad prompt and progressively refining it, educators can develop a practical sense of how to guide the AI to produce content that aligns with their teaching goals.

Try it yourself: Context Variation in AI Prompts

- **Start Simple**: Craft a basic prompt related to your subject with minimal context, such as "Describe polynomials."

- **Gradual Enhancement**: Incrementally add more context to the prompt. First, specify the grade level like "Describe polynomials for 8th-grade students." Then, add more details, such as "Explain how to solve polynomial equations in 8th-grade algebra." Get more specific by clarifying your purpose, such as "Explain how to solve polynomial equations in 8th-grade algebra and what they are used for in the real world."

- **Observe and Note**: With each addition, observe how the AI's responses change. Take notes on the level of detail, relevance, and accuracy.

- **Reflect**: After experimenting, reflect on how varying amounts of context impacted the AI's understanding and usefulness of the response.

- **Apply**: Use these insights to refine future prompts for your classroom use.

This exercise shifts from refining prompts to reimagining them, helping educators explore how context shapes AI's responses. Starting with a general, educator-centered prompt offers insight into how AI interprets tasks. Reframing prompts as student questions encourages the AI to address curiosity and common misunderstandings, making outputs more learner-focused.

This role reversal highlights how prompts tailored to different audiences align better with educational objectives. It helps educators identify gaps in materials and refine prompts to produce more comprehensive, actionable content that bridges teaching goals with student needs.

Try it yourself: Role-Reversal Exercise

Take a General Prompt: Begin with a basic directive, such as "Design a lesson on photosynthesis."

Reframe as a Student Question: Shift the perspective: "What questions might a student ask about photosynthesis?"

Observe the Shift: Analyze how changing the perspective enhances the depth and relevance of AI-generated content.

Concluding Insight

These exercises showcase how experimenting with prompts empowers educators to refine their approach to AI tools. By iteratively improving clarity, specificity, and context, educators can craft prompts that transform AI into an indispensable ally in the classroom, capable of delivering engaging, meaningful, and student-centered content.

Common Pitfalls to Avoid

Crafting effective AI prompts requires precision and clarity, but even experienced educators can stumble into common pitfalls when incorporating context. These missteps can undermine the effectiveness of a prompt, producing responses that are either irrelevant, overly complicated, or skewed away from the intended goal. Understanding these potential issues is critical for refining your approach and ensuring that prompts align with your educational objectives.

When designing prompts, it's essential to strike a balance between providing sufficient guidance and leaving room for the AI to generate creative and flexible responses. Overloading prompts with unnecessary details or omitting key elements can lead to subpar outcomes. The following list identifies frequent mistakes and offers insight into how to avoid them, so you can maximize the utility of AI in your teaching practice.

- *Overstuffing Prompts:* Too much detail can confuse the AI, leading to convoluted responses.
- *Misleading Context:* Inaccurate or ambiguous context can skew the AI's output.

- *Neglecting Relevance:* Ensure the context aligns closely with the educational goal to avoid irrelevant outputs.

Each of these pitfalls illustrates how missteps in prompt crafting can diminish the quality and relevance of AI-generated responses. Overstuffing prompts with excessive details might seem like a way to guide the AI more precisely, but it often results in convoluted answers that are difficult to interpret or apply. This overload of information can obscure the main task and reduce the prompt's effectiveness.

Similarly, misleading or ambiguous context can confuse the AI, steering it toward producing responses that are off-topic or unhelpful. Ensuring clarity in the information provided helps maintain focus on the educational objective. Finally, neglecting relevance by failing to align the context with specific goals risks generating responses that, while accurate, do not serve the intended purpose. Avoiding these pitfalls ensures that your prompts remain targeted, clear, and useful for both educators and students.

The Power of Context in Prompt Engineering

Context transforms AI from a generic tool into a dynamic partner in education. By embedding appropriate details into prompts, educators can guide AI systems to deliver insightful, accurate, and engaging responses. Through experimentation and refinement, teachers can master the art of context-driven prompt engineering, unlocking the full potential of AI to create meaningful learning experiences. Keep refining your prompts, and watch as your classrooms come alive with innovative, tailored content.

Chapter 5: Understanding Biases and Limitations

AI bias occurs when machine learning models, trained on historical data, produce skewed or unfair outcomes. For educators, recognizing and addressing bias is crucial to maintaining fairness, inclusivity, and accuracy in the classroom. Biases in AI-generated content can inadvertently perpetuate stereotypes or reinforce systemic inequities, influencing assessments, teaching materials, and even resource allocation. Educators must approach AI tools critically, understanding the inherent limitations and taking steps to mitigate bias to create a positive and equitable learning environment.

Biases can emerge from the data that trains AI models, which often reflects societal inequities. For instance, biased language in training data can lead to discriminatory outputs. This is particularly concerning in educational contexts, where students from diverse backgrounds rely on equitable tools to enhance their learning. Addressing bias is not only about improving AI's accuracy but also about ensuring that educational practices promote inclusivity and respect for all perspectives.

Recognizing Bias in AI Responses

Bias in AI-generated content can manifest in subtle or overt ways across various educational contexts, from historical narratives to cultural roles and literature analysis. When prompts fail to specify the need for balanced perspectives, AI outputs often reflect dominant narratives, ignoring less-documented or marginalized viewpoints.

This can inadvertently skew students' understanding, reinforce stereotypes, and diminish the inclusivity of educational content. Below, we explore how bias manifests in three key areas and the steps educators can take to address it effectively.

Historical Interpretation

Historical narratives are inherently shaped by the perspectives of those who document them, often prioritizing dominant viewpoints while marginalizing others. When AI systems trained on such data are used in educational contexts, they risk perpetuating these biases, offering incomplete or skewed interpretations of historical events. For instance, a prompt like :

```
Explain the causes of [historical event]
```

may lead the AI to present a surface-level analysis rooted in widely accepted narratives, potentially neglecting the perspectives or contributions of underrepresented groups. This lack of depth not only reinforces existing misconceptions but also limits students' ability to critically engage with the multifaceted nature of history, underscoring the importance of carefully crafting prompts to guide the AI toward more inclusive and balanced responses.

Consider a scenario involving the Industrial Revolution. A general AI response might emphasize technological advancements and economic growth in Europe while omitting the perspectives of colonized nations, whose resources fueled this transformation. Such omissions narrow students' understanding of the interconnected global impact of this era.

When discussing complex historical events like the Industrial Revolution, it's crucial to ensure that multiple perspectives are considered. This prevents the AI from defaulting to oversimplified or one-sided narratives that may omit the experiences of less dominant groups. A prompt like:

> What were the impacts of the Industrial Revolution?

might yield a response focused solely on industrialized nations, neglecting the profound effects on colonized regions or marginalized populations. By refining the prompt to:

> Discuss the impacts of the Industrial Revolution from the perspectives of industrialized nations and colonized regions, highlighting both economic benefits and social challenges.

educators can guide the AI to provide a more nuanced and inclusive analysis, fostering a deeper understanding of the interconnected global dynamics of historical events.Moreover, critical evaluation of AI responses is essential. Educators can guide students in questioning the sources and biases inherent in the AI's outputs, fostering a more analytical approach to historical study. By emphasizing multiple viewpoints, educators enable learners to engage with history as a multidimensional and contested field.

Cultural Analysis

Cultural analysis is another area where biases in AI

responses can emerge, particularly when addressing topics like gender roles, traditions, or societal structures. For instance, a prompt such as:

> Describe the role of women in [specific culture or era].

may lead to an AI-generated response that perpetuates stereotypes or overgeneralization. These biases often stem from limited or one-sided training data, which fails to capture the diversity and evolution within cultures.

Take the example of discussing women's roles in ancient Greece. A generic AI response might focus solely on domestic roles or the constraints placed on women, neglecting the varied experiences of women across city-states or the contributions of figures like the poet Sappho. Such oversimplifications not only misrepresent the culture but also reinforce narrow stereotypes.

When exploring cultural roles in history, it's vital to craft prompts that encourage depth and highlight diversity, ensuring students develop a richer and more nuanced understanding of the subject. A well-constructed prompt such as:

> Analyze the roles and contributions of women in ancient Greece, comparing their experiences in different city-states such as Athens and Sparta, and including perspectives from art, literature, and archaeology.

sets clear expectations for the AI to explore the topic comprehensively. By specifying the inclusion of diverse sources—art, literature, and archaeology—and contrasting the societal norms of Athens and Sparta, the prompt guides the AI to provide a well-rounded and insightful response. This approach not only minimizes the risk of bias or oversimplification but also fosters a deeper appreciation for the complexities and variations within cultural and historical contexts. Such prompts are invaluable for creating educational materials that celebrate diversity and engage learners with critical analysis.

Additionally, educators should encourage students to evaluate the sources and assumptions in AI-generated analyses. By fostering discussions around cultural complexity and avoiding monolithic representations, teachers can cultivate a deeper appreciation for the richness and variability of human societies.

Literature Discussions

Literary analysis often serves as a window into societal attitudes and values, but biases in AI-generated responses can distort this perspective. A question like:

```
Discuss the portrayal of [group] in [text].
```

may yield an overly simplistic interpretation, rooted in prevailing stereotypes found in the training data. Such responses can obscure the text's deeper themes and perpetuate reductive views of characters or groups.

For instance, analyzing the representation of indigenous characters in a classic novel might lead the AI to highlight only tropes of victimhood or mysticism, ignoring instances of agency, resilience, or cultural richness. These biased portrayals can hinder students from engaging critically with the text and appreciating its complexities.

To ensure a balanced and thoughtful exploration of sensitive topics, educators can design prompts that emphasize complexity, encourage critical thinking, and incorporate multiple perspectives. For instance, rather than posing a straightforward question like:

```
How does [author] portray indigenous characters in [work]?
```

educators can reframe prompts to encourage deeper analysis and engagement. A well-crafted prompt moves beyond surface-level understanding, guiding both the AI and students to explore themes, context, and broader implications for a richer learning experience:

```
Examine the portrayal of indigenous characters in [work], considering themes of agency, resilience, and cultural identity, and comparing these depictions to the historical contexts of the era.
```

This expanded prompt provides clear guidance, prompting the AI to analyze nuanced elements such as character agency and resilience while also connecting literary representations to historical realities. By doing so, educators can ensure that the AI delivers responses that are not only academically rigorous but also culturally sensitive and inclusive, helping students engage with literature in a way that fosters empathy, critical reflection, and an appreciation for diverse perspectives. Such prompts equip learners with a deeper understanding of the material and its broader implications, encouraging them to think beyond the text and consider its relevance to societal narratives and historical truths.

Educators should integrate supplementary materials to provide broader contexts for literary discussions. This might include historical documents, critical essays, or creative works by authors from the group being represented. By broadening the lens through which students approach literature, teachers can ensure that analysis remains inclusive and thought-provoking.

By addressing bias in these three areas—historical interpretation, cultural analysis, and literature discussions—educators can guide students toward more comprehensive and equitable understandings. Thoughtful prompt design and critical engagement with AI responses not only enrich educational content but also foster a learning environment that values diversity, challenges stereotypes, and encourages open-minded inquiry.

Rephrasing Prompts to Mitigate Bias

The way prompts are constructed profoundly impacts the quality, relevance, and inclusivity of AI responses. Thoughtful prompt design is essential for mitigating bias and ensuring a balanced perspective. Incorporating inclusive language and framing questions to explore diverse viewpoints help guide the AI toward producing well-rounded outputs that reflect a wider range of experiences and interpretations. For example, a prompt such as:

> **Discuss the various cultural, economic, and social factors that contributed to [historical event], highlighting multiple viewpoints and scholarly debates.**

shifts the focus from a single narrative to a broader exploration of factors and interpretations. This approach encourages the AI to delve into different contexts and perspectives, creating responses that are more nuanced, equitable, and reflective of the complexity of the subject. Such prompts not only enhance the depth of the content but also contribute to fostering a more inclusive learning environment, enabling students to engage with diverse ideas and think critically about historical narratives.

Case Study:

Framing Prompts for Inclusivity in Historical Analysis

I. **Scenario:** An educator teaching about the American Revolution wants students to examine multiple perspectives on the war. They aim to highlight the experiences of various groups, including colonists, Native Americans, and enslaved individuals, to provide a comprehensive understanding of the historical event.

II. **Initial Prompt:** *"Describe the causes of the American Revolution."*

A. *Outcome:* The AI generates a general response, focusing primarily on taxation without representation and colonial resistance to British policies. The perspectives of marginalized groups are overlooked.

B. *Analysis:* This prompt lacks direction for the AI to consider diverse viewpoints, resulting in a surface-level, one-dimensional narrative.

III. **Revised Prompt:** "*Discuss the causes of the American Revolution from the perspectives of colonists, Native Americans, and enslaved individuals. Include how their experiences and motivations differed.*"

A. *Outcome:* The AI provides a more detailed response, incorporating the economic and political grievances of the colonists, the impact of the conflict on Native American tribes, and the role of enslaved individuals, including their struggles and contributions.

B. *Analysis:* This refined prompt guides the AI to consider multiple viewpoints, creating a more inclusive and comprehensive analysis of the historical event.

IV. **Final Prompt:** "*Examine the causes of the American Revolution by analyzing the perspectives of three distinct groups: colonists advocating independence, Native Americans caught in the conflict, and enslaved individuals seeking freedom. Highlight how their motivations and experiences shaped the broader narrative, and include scholarly debates on these perspectives.*"

A. *Outcome:* The AI generates an in-depth response, including specific examples such as the role of the Iroquois

Confederacy, the petitions for freedom by enslaved individuals, and the philosophical writings that inspired colonial leaders. The response also incorporates academic interpretations, enriching the discussion.

B. *Analysis:* The final prompt not only ensures inclusivity but also elevates the complexity of the response by integrating scholarly debates and specific examples. This fosters critical thinking and engagement with diverse historical narratives.

V. **Key Takeaways:**

A. *Initial Prompt:* Broad and generic, leading to a narrow and incomplete perspective.

B. *Revised Prompt:* Adds direction by specifying perspectives, resulting in a more inclusive response.

C. *Final Prompt*: Combines inclusivity with scholarly depth, encouraging critical thinking and nuanced understanding.

D. *Lesson:* Thoughtfully framed prompts guide AI to produce content that reflects diverse viewpoints, creating richer and more equitable learning experiences.

By iterating on prompt design, educators can use AI as a powerful tool to enrich classroom discussions and support a deeper, more inclusive exploration of historical topics.

Potential Consequences of Biased AI Content

The integration of AI into educational settings has the potential to revolutionize learning, but it also brings significant risks if biases in AI-generated content go unchecked. When biases are present, they can shape students' understanding, skew decision-making processes, and perpetuate systemic inequities. These effects can undermine the transformative promise of AI in education, making it imperative for educators to recognize and address the potential consequences of biased content. From amplifying existing disparities to ethical concerns about representation and fairness, understanding these risks is essential for fostering an equitable and inclusive educational environment.

The effects of biased AI responses can extend far beyond classroom discussions. Consequences include:

- *Reinforcement of Educational Disparities:* AI systems trained on biased data may perpetuate systemic inequities, disadvantaging marginalized groups.
- *Unfair Resource Allocation:* When AI-powered tools influence funding or support decisions, biases in training data can lead to unequal resource distribution.
- *Undermining the Learning Process:* Over-reliance on biased AI-generated content can hinder critical thinking and authentic student engagement.
- *Ethical Implications:* Biased AI tools can perpetuate harmful stereotypes, contradicting the goals of equity and inclusivity in education.

The potential consequences of biased AI content underscore the critical need for vigilance and responsibility in its use within educational settings. By carefully refining prompts, assessing AI outputs, and advocating for diverse and representative training data, educators can mitigate these risks and leverage AI as a tool for equitable and inclusive learning. Proactive measures to address biases ensure that AI supports a fairer educational landscape, fostering critical thinking, authenticity, and equal opportunities for all students. As AI becomes more deeply integrated into classrooms, its success will depend not only on technological innovation but also on the ethical commitment of educators to promote fairness and equity.

Try it yourself: Identify Bias in AI Responses

1. Craft a prompt on a complex topic such as cultural analysis or historical interpretation.
2. Observe the AI's response and analyze it for signs of bias, such as *one-sided narratives* or *stereotypical language*.
3. Then revise the prompt to request multiple perspectives or diverse examples.
4. Compare the original and revised outputs to understand how prompt structure affects bias.

Why this is beneficial:

This exercise trains you to recognize subtle biases in AI responses and refine prompts to promote inclusivity. By actively engaging with AI-generated content, you can develop a sharper eye for ensuring balanced and fair responses.

Try it yourself: Eliminating Bias Through Reframing

1. Take an existing prompt and run it into your preferred AI (I like ChatGPT for this). See what you get.
2. Then use the prompt: "What are the biases in your answer?" Review the information provided.
3. Then use the prompt: "How could I reframe the prompt to eliminate bias?" Review the suggestions.
4. Rewrite the prompt with the intent of eliminating bias and try this process again.

Why this is beneficial:

This activity is a great way to dive into how AI responds and uncover any hidden biases in its answers. Through getting the AI to help you analyze it's response and refine your own question, you'll not only create more inclusive and fair prompts but also sharpen your ability to spot and address bias in real time. It's an iterative, hands-on way to ensure your prompts are as thoughtful and impactful as they can be, all while building your awareness of how language shapes AI outputs.

Bias in AI is an ongoing challenge that educators must address to ensure equitable and accurate learning experiences. Through critical evaluation and thoughtful prompt engineering, educators can guide AI tools to generate content that supports inclusivity, fosters critical thinking, and enhances educational outcomes. The strategies in this chapter empower educators to use AI responsibly, leveraging its potential while safeguarding against its limitations. Stay proactive, experiment with new approaches, and continue refining your skills to create a learning environment where every student feels seen and respected.

Chapter 6: Understanding Effective Prompts in AI Engineering

Crafting effective AI prompts is both an art and a science. It's not just about what you ask but how you ask it. Think of prompts as a bridge between your educational goals and the AI's capabilities. A well-structured prompt can transform AI into a powerful teaching tool, delivering precise, insightful, and actionable content. This chapter delves into the core principles of prompt design, offering you the tools to create prompts that maximize AI's potential while aligning closely with your classroom needs. Whether you're guiding students through complex historical events or simplifying scientific concepts, the key lies in the thoughtful formulation of your prompts.

Prompt engineering requires a deep understanding of how language shapes AI responses. By analyzing and refining prompts, educators can uncover what works and what doesn't, ensuring every query yields useful and relevant outcomes. This chapter isn't just about theoretical knowledge; it's about empowering you to craft prompts that guide AI toward generating exactly what you need—whether it's a lesson plan, an assessment, or an engaging activity for your students.

Reviewing and Analyzing Prompts

To understand prompt engineering, we must examine both effective and ineffective examples. Each prompt, whether successful or not, offers insights into the mechanics of AI communication.

Effective prompts lead to clear, targeted responses, while ineffective ones highlight gaps or ambiguities in phrasing. By evaluating prompts with a critical eye, we can refine our approach and ensure better results in the future.

Broad prompts provide a general overview, laying the groundwork for more detailed inquiry. This approach is especially valuable when you're unfamiliar with the subject matter or need a foundational understanding before diving deeper. However, while this initial exploration sets the stage, it may not deliver the precision or depth required for targeted educational goals. Starting with a broad science prompt such as:

```
Explain photosynthesis.
```

This serves as a solid starting point for general information but lacks depth and specificity. A teacher researching Game Theory or another unfamiliar topic might use such a prompt to initiate an exploratory conversation with AI. However, the results often resemble a Wikipedia-style overview—helpful but not tailored to immediate classroom needs.

Refining broad prompts into detailed and purposeful ones is essential for extracting actionable results from AI. While general prompts provide a foundation, they often lack the precision needed for practical classroom applications. By building on the previous example of refining a basic query into a detailed response about photosynthesis, we see the importance of adding specificity, such as purpose, audience, and desired outcomes. A refined prompt like:

Explain the light-dependent reactions of photosynthesis for a 10th-grade biology class, and create a 5-question multiple-choice quiz to assess student understanding.

ensures the output is not only focused but also tailored to lesson planning and assessment needs. This process enables educators to craft prompts that yield precise, relevant, and classroom-ready resources.Such adjustments transform broad queries into specific, goal-oriented prompts. Educators should always consider three key factors:

1. *Goal:* What specific information or outcome do you want from the AI?
2. *Context:* Who is the audience, and what level of understanding should the response target?
3. *Intent:* How will the AI's response be used—e.g., for lesson planning, quizzes, or student activities?

By addressing these questions, you ensure your prompts are clear, actionable, and aligned with your educational goals. Clear prompts guide the AI to produce relevant responses, while actionable prompts ensure the output can be directly applied to lesson planning, assessments, or class materials. Aligning prompts with objectives supports broader curriculum goals, such as enhancing comprehension or meeting specific learning outcomes. Refining your prompts maximizes AI's efficiency, saving time and fostering productive interactions that fully leverage its capabilities for your classroom needs.

Components of Effective Prompts

Creating effective prompts is the key to unlocking AI's full potential in education. Think of a well-crafted prompt as a map—it directs the AI to deliver responses that are accurate, relevant, and perfectly suited to your needs. By zeroing in on key elements like specificity, context, clarity, and task formulation, you can get the most out of AI tools while saving yourself time and effort. Let's take a closer look at these building blocks and explore how they can transform the way you use AI.

- *Specificity:* Clearly define the subject and scope of the response to avoid overly broad or irrelevant outputs.
- *Context:* Provide details about the intended audience, such as grade level or prior knowledge, to ensure the response is appropriately tailored.
- *Clarity:* Use straightforward language to prevent misinterpretation.
- *Task Formulation:* Clearly outline the desired outcome, such as creating a quiz, developing a lesson plan, or summarizing a concept.

Applying the Elements Effectively

Understanding how the elements of specificity, context, clarity, and task formulation come together is essential for crafting effective prompts. When educators thoughtfully integrate these components, they transform broad, unfocused queries into precise and actionable tools that enhance learning outcomes. Let's examine how these elements work in practice, using real-world examples to highlight their value.

Broad Prompt: Describe World War II.

- *Why It Falls Short:* The lack of specificity results in a broad, surface-level overview that may touch on causes, events, and impacts but lacks depth or focus. This type of prompt is functional for general exploration but doesn't align well with specific teaching goals.
- *Improvement Opportunity:* Identify the aspect of World War II you wish to explore. Is it social impacts, military strategies, or political consequences? Define a more specific scope to narrow the response.

Refined Prompt: Summarize the impact of propaganda during World War II on public morale in England and Germany, and suggest three sources for further student research.

- *What This Adds:* This iteration focuses on a specific topic—propaganda—and limits the scope to two countries, making the prompt targeted and actionable. Including a request for research sources provides additional utility for classroom preparation.

- *Improvement Opportunity:* While more focused, the prompt could benefit from added instructional purpose or format to guide how the response should be delivered. For example, should the output be a summary, presentation, or discussion guide?

Advanced Refinement:

> Develop a five-slide presentation on the impact of World War II propaganda on public morale in England and Germany, including visual examples and a discussion question for each slide.

- *Why It Works:* This version provides clear, detailed instructions with an intended output format (presentation), a targeted topic (propaganda's impact), and supplementary elements (visual examples and discussion questions). These details help the AI generate a specific, ready-to-use resource.
- *Final Consideration:* Ensure alignment with your teaching objectives. This prompt assumes the need for engaging, multimedia content. Adjust instructions as necessary for different learning styles or outcomes.

Why This Matters

Mastering the art of prompt crafting is an essential skill for educators in the AI age. By thoughtfully incorporating specificity, context, clarity, and task formulation, teachers can ensure their prompts elicit responses that are not only informative but also directly applicable to their classroom needs. Each of these components plays a crucial role in aligning AI outputs with educational objectives, turning AI into a versatile tool that enhances both student engagement and learning outcomes.

As demonstrated in the comparison of prompts, even small adjustments can significantly impact the quality and relevance of AI-generated responses. A vague prompt may provide a functional but unfocused answer, while refined prompts evolve into powerful tools that align precisely with teaching goals. These refinements not only save educators valuable preparation time but also encourage deeper student engagement by fostering critical thinking and sparking meaningful classroom discussions.

Effective prompt crafting is more than a skill—it's a gateway to unlocking AI's full potential in education. By transforming vague queries into actionable tools, educators can create learning experiences that are tailored, interactive, and impactful. With consistent practice and attention to these key elements, educators can make AI an indispensable ally in delivering precise, meaningful outputs that enhance the learning environment.

By understanding and applying these principles, educators can unlock AI's full potential, transforming it into an invaluable partner in creating engaging, relevant, and impactful educational experiences. Whether you're developing lesson plans or brainstorming student activities, effective prompts bridge the gap between AI's capabilities and your classroom needs. Let's put these insights into practice and watch as your teaching tools become sharper, smarter, and more aligned with your goals.

Try it yourself: Promptology 101

Identify a topic you're currently teaching and create a basic prompt to explore it. Before running the prompt, ask yourself:

- ***Goal:*** Does the prompt clearly state the desired outcome or purpose?
- ***Context:*** Have you included relevant details about the audience (e.g., grade level, subject area) and the intended application (e.g., quiz questions, lesson plans)?
- ***Clarity and Specificity:*** Is the language straightforward and detailed enough to avoid vague or overly broad responses?

Refine the prompt by addressing any missing elements. Then, run both the original and revised versions through your preferred AI tool. Analyze the results by comparing the outputs for relevance, depth, and usability in your classroom. Reflect on the improvements and identify strategies to apply these refinements to future prompts.

Generative Content Framework

The Generative Content Framework is best suited for queries that involve the creation of educational content, such as

Generative Content Flowchart

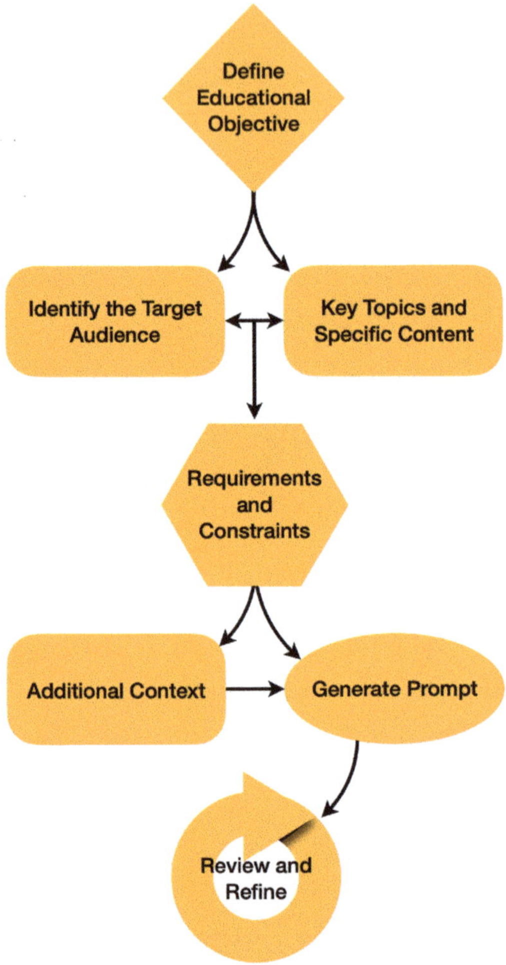

Scan to access the online image of the Generative Content Flowchart

I am submitting a personal narrative essay as part of my college application, and I need guidance to improve it. Please act as a writing tutor and provide detailed, sentence-level feedback on the essay. Your analysis should focus on three key areas: clarity of writing, use of narrative structure, and depth of reflection. Highlight what is working well and pinpoint areas that need improvement, but refrain from rewriting or rephrasing any sentences. Instead, offer actionable advice for enhancing each part of the essay. The essay is as follows:

lesson plans, explanations, and learning materials. This includes scenarios where educators need to develop comprehensive educational content packages aligned with specific educational objectives and standards. Additionally, it is ideal for crafting prompts that guide AI algorithms to generate content tailored to the needs of a particular target audience, such as students in a specific grade level or subject area.

Leveraging AI algorithms to generate educational content, the Content Generation Algorithm emphasizes a step-by-step process. This process begins with defining the educational objective and identifying the target audience, followed by specifying the content type, outlining the key topics or concepts, and providing any unique requirements or constraints. Additionally, it underscores the importance of providing additional context, testing and refining the prompt, and gathering reference materials. The formula is designed to streamline and enhance the content creation process, ensuring that the generated content aligns with curriculum standards and educational objectives. By applying the Generative Content Frameworks formula, weaker prompts can be refined to include specific educational objectives, the intended target audience, and the key topics or concepts to be covered. This transformation ensures that the AI-generated content is tailored to meet the specific educational needs and standards, ultimately enhancing its effectiveness and relevance.

By providing clear and specific instructions for the AI algorithm to follow, it guides the AI in understanding the educational objective, the target audience, and the content type, enabling the AI to generate educational content that is aligned with specific learning objectives and standards. By inputting a well-crafted prompt into the AI system, the AI uses this information to perform the analysis and generate personalized recommendations, ensuring that the content generated meets the necessary educational criteria and enhances the learning experience for students.

Using the Generative Content Flowchart

Let's take a moment to break down the aspects of this flowchart. When working to develop content, be that a short reading or a lesson plan, it's important to begin with your objective. Think of it like backwards planning in prompt engineering. At the beginning of your prompt, you should define your educational objective, let the AI know what you want it to create.

The second step can vary based on what you're prompting the AI to do. Either you can identify the target audience—I recommend stating the age and level of the people for which you're generating content—or you can state the key topics and specific content such as in mindfulness exercises for stressed out professionals. Either works to help add specificity, but better still, in most cases, is to use both.

Finally, and possibly most importantly, you need to give some direction to the AI to keep it from going off the rails. Explain what your requirements are and give it some constraints to work within. I find this step to be the most important in generating effective and useful content. If you look in the sample prompt on the Generative Content Flowchart you'll see that I provided the AI with the constraint that it should provide a "granular line-by-line analysis" to "provide sentence-level feedback" from and it should not "rewrite or rephrase any specific sentence." These constraints made it clear to the AI that I was looking for feedback on the essays, not a simple rewrite.

Exploring the Sample Prompt

I actually used this prompt in my senior English classes to help my students more from a rough draft to a second draft of their essays. They found the process very helpful and informative, and it literally took me moments to get all students individualized, granular feedback that would have taken me days to do before. This doesn't replace the teacher's eyes on the papers, nor should it, but it will help them hand in a better draft when they do.

For this prompt, I needed them to include additional content, which you can see is an optional step in the framework, just before prompt generation. In this case, the additional concept was the rough draft of their essay, other times it's been excerpts from a story or even a lesson that I wrote or a project I'd developed. It all depends on what your initial goal is.

You'll notice that at the bottom of the framework, after generating the prompt—which simply means plugging it into the AI system's chat box and running it—is the review and refine portion of the program. Some prompts, when you use this model, work perfectly, other work only passably. In the latter case, you may need to use the iterative process to zone in on what you wanted. This is not a failing, not of you or the system. It's part of the learning process. We try, we struggle, we try some more, just like we tell our students. It's the iterative process of prompt engineering that makes the process so interesting—and daunting—to people. However, I've noticed that the more we develop prompts, the smoother the process becomes and the less we need to iterate our prompts.

Curricular Alignment Flowchart

The Curricular Alignment Flowchart is best suited for times when you're looking to ensure your content aligns with curriculum standards and learning objectives or that your school's curriculum aligns with the state standards. This includes scenarios where educators need to analyze and align curriculum standards with specific learning objectives, ensuring that their lessons are in compliance with the curriculum standards. Additionally, it is ideal for crafting prompts that guide AI algorithms to generate reports that highlight the alignment and any potential gaps between the curriculum standards and learning objectives or state standards.

Though a step-by-step process, starting with defining the curriculum standards and identifying the specific learning objectives, the prompt should determine the educational context, craft a clear and specific prompt, and input the prompt into the AI system. Once the AI generates a report, the user should review and interpret the report, make necessary adjustments, and periodically reassess and

Curricular Alignment Flowchart

Scan to access the online image of the Curricular Alignment Flowchart

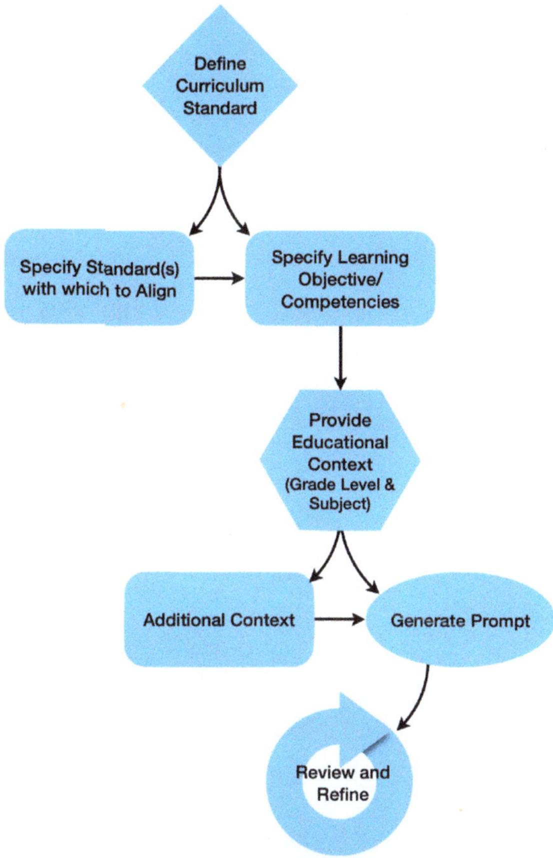

Using AI algorithms, analyze the 6th-grade mathematics curriculum standards for the state of Massachusetts. Break these standards down into specific learning objectives or competencies. Next, compare these results to the attached curriculum map to identify alignment and potential gaps. Include the following in your AI-generated report: (1) a clear summary of how the standards align with the curriculum map, (2) specific areas where gaps exist, and (3) recommendations for addressing those gaps. Ensure the report is structured to highlight opportunities for lesson plan adjustments, activities, or assessments to improve alignment.

update the curriculum alignment. This formula is essential for educators to maintain educational standards and ensure that their lessons are aligned with curricular standards. By applying the Curricular Alignment Flowchart, such a prompt can be refined to include specific curriculum standards, the intended learning objectives, and the educational context. This transformation ensures that the AI-generated report is tailored to meet the specific educational needs and standards, ultimately enhancing its effectiveness and relevance.

The Curricular Alignment Flowchart communicates with the AI by providing clear and specific instructions for the AI algorithm to follow. It guides the AI in understanding the curriculum standards, the learning objectives, and the educational context, enabling the AI to generate a report that highlights the alignment and any potential gaps. By inputting a well-crafted prompt into the AI system, the AI uses this information to perform the analysis and generate personalized recommendations, ensuring that the curriculum is in compliance with the necessary educational criteria and enhances the learning experience for students.

Using the Curricular Alignment Flowchart

Using the flowchart effectively involves a step-by-step process to ensure your prompts are clear, focused, and actionable. First, start by defining your educational objective. Be specific about what you want from the AI, whether it's aligning curriculum standards, generating activities, or creating assessments. For example, "*Identify areas of alignment between 6th-grade Massachusetts music standards and my school's current general music curriculum.*" This type of goal provides clarity for both you and the AI, ensuring the response directly addresses your needs by narrowing its focus to a manageable task.

Be sure you have identified your target audience. Determine who will benefit from the AI's response, such as students, teachers, or administrators. Specify the grade level, subject, or group, so the AI can tailor its response appropriately. For instance, if the task is for 6th-grade general music students, the AI knows to avoid advanced theoretical concepts that are beyond their understanding. This step ensures that the output is not only relevant but also accessible to the intended audience.

After that, focus on key topics and content. Highlight the specific elements you want the AI to prioritize, such as rhythm, melody, or music theory. For example, "*Focus on rhythm, melody, and introductory music theory to ensure alignment with standards.*" By directing the AI toward these areas, you guide it to avoid irrelevant tangents, improving the quality and usefulness of the results.

Outline requirements and constraints next. Constraints, such as adhering to state guidelines or referencing specific learning outcomes, help the AI remain within the intended scope. For example, "*Limit the analysis to standards applicable to beginner-level music students and include objectives that can be assessed through performance or composition activities.*" These constraints ensure the AI doesn't provide information that is too broad or inappropriate for the task at hand.

Finally, add any necessary context, such as resources or background information. Providing additional materials, like a curriculum map or sample lesson plan, gives the AI the reference points needed to deliver accurate results. For instance, "*Use the attached curriculum map as a reference when identifying gaps in rhythm and melody instruction.*" This ensures the AI aligns its response with the specific context of your request, resulting in a more relevant and actionable output.

By following these steps, you ensure that the AI's responses are precise and tailored to your educational objectives. The examples show how clearly defined goals and structured prompts improve outcomes, and the explanations emphasize why each element is essential. This approach not only saves time but also enhances the value AI brings to your tasks, helping you create meaningful, impactful resources.

Exploring the Sample Prompt

The sample prompt effectively targets the Massachusetts 6th-grade mathematics standards, providing a defined starting point for analysis. By clearly naming these standards, the AI is guided to operate within a specific scope, ensuring relevance and precision. It also instructs the AI to break these standards into specific learning objectives, encouraging a granular approach to understanding the curriculum. This focus allows the AI to generate detailed insights that are both actionable and aligned with the educational goals.

The inclusion of the grade level and subject provides essential context, ensuring the results are tailored to the intended audience. The prompt further enhances its effectiveness by specifying tasks such as comparing the standards to a curriculum map and identifying gaps. These specific directives guide the AI to produce targeted, actionable outcomes. Additionally, the request for a structured report with actionable recommendations elevates the prompt's utility, providing educators with practical tools for addressing curriculum gaps and optimizing alignment.

By emphasizing the importance of gap analysis and actionable recommendations, the prompt supports a review and refinement process. Although the iterative reassessment phase is not explicitly included in the initial prompt, the recommendations provided by the AI lay the groundwork for ongoing evaluation and improvement. This iterative approach ensures that the curriculum remains dynamic and responsive to educational needs.

This structured design aligns closely with the flowchart's steps, integrating elements like clear goals, thoughtful task formulation, and actionable outcomes. The prompt not only ensures compliance with standards but also equips educators with strategies to enhance student learning. By moving from general analysis to targeted and practical solutions, the prompt maximizes the value of AI in an educational context.

Assessment Item Generation Flowchart

Scan to access the online image of the Assessment Item Generation Flowchart

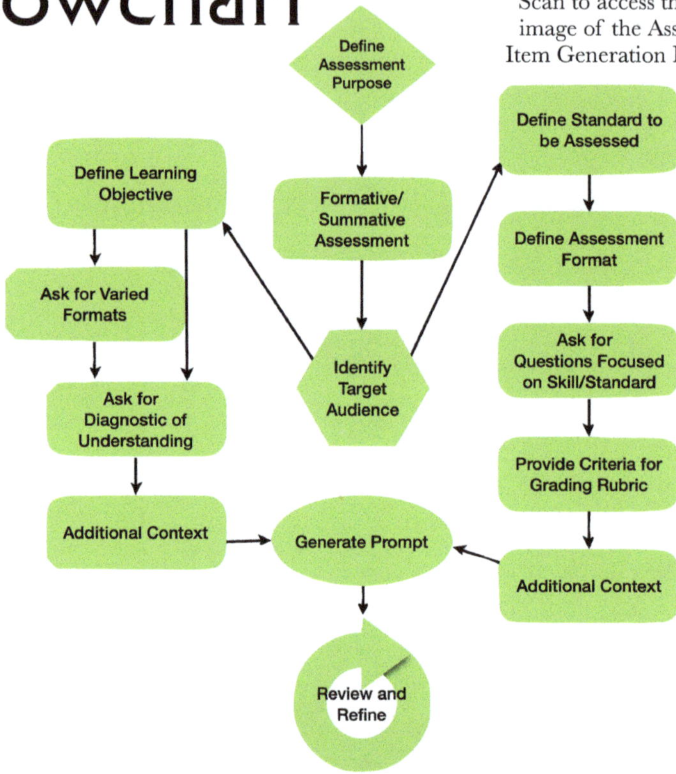

I just finished teaching my 6th-grade social studies class a unit on European countries and capitals, and I need a summative assessment. The test should evaluate students' ability to match countries to their capitals and identify countries based on their capitals. I'd like 20 questions, evenly split between multiple-choice and fill-in-the-blank formats, appropriate for 6th graders. Please include an answer key and ensure the questions align with the unit content.

Assessment Item Generation Flowchart

This formula is focused on using AI to create a variety of assessment items, including quizzes, tests, and assignments. It involves defining the assessment purpose, identifying the target audience, determining the assessment type and format, and specifying the assessment content and learning objectives. The prompt should be crafted to create a clear and specific prompt that instructs the AI algorithm on what assessment items to generate. It should also consider the specific goals and objectives of each assessment type. An example prompt is provided, guiding the AI algorithm to generate a 9th-grade history quiz on the topic of 'World War II,' consisting of multiple-choice and short answer questions, and ensuring that the questions are well-balanced in terms of difficulty and aligned with the curriculum standards.

AI algorithms can help educators make data-driven decisions about curriculum alignment by providing a systematic and objective analysis of how well educational content aligns with curriculum standards and learning objectives. By utilizing AI algorithms, educators can receive detailed reports that highlight the alignment and any potential gaps between the curriculum standards and the learning objectives. This data-driven approach allows educators to identify areas that may need adjustments and make necessary modifications to ensure that their educational content remains in compliance with the curriculum standards. Additionally, AI algorithms can provide personalized recommendations on how to adapt instruction to meet individual student needs, ultimately enhancing the learning experience for students. This data-driven approach ensures that educators have the necessary insights to maintain educational standards and make informed decisions about curriculum alignment.

Using the Assessment Item Generation Flowchart

When using this flowchart, you should begin with identifying the reason for the assessment, if you plan for it to be a formative or summative assessment, and your target audience. For instance you could be trying to assess the knowledge of geometric shapes in a formative assessment for an honors-level sophomore geometry class. Starting your prompt with that information allows the AI to tailor the assessment item to be tailored to your needs. After the initial information, you need to make a choice, are you assessing based on your learning objectives or based on state standards.

If your assessment items are being generated for learning objectives, which is often the case with formative assessments, then you're going to need to determine if you want a varied format in the assessment items or a single type of assessment. Are you looking for a single project or for a mix of multiple-choice, fill-ins, and short answer questions or something in between? Once that is decided, you need to determine your diagnostic of understanding. What, specifically, do you want to be sure the students understand when they complete the assessment. Finally, add in any additional context that may help, then run your prompt and generate the AI's response.

Alternately, if you're looking for a standards based assessment, specify the standard or standards you're assessing, then define your assessment format—again, this could be a test, quiz, question set, or some other assessment. Then you want to ask for questions on specific skills or standards to drill down on the exactly what you want to assess. At this point you want to identify and include your grading criteria, this could be accomplished by uploading a rubric for any written work being assessed. The AI can than incorporate that into the assessment questions. If you do not have one, you can request the AI to create a rubric for any written portions, but I would recommend you provide it with your rubric criteria at the minimum. Then, once again, add in any additional content that could be beneficial and run your prompt.

Regardless of the path you take, remember that you should review the work produced by the AI in order to assure it is accurate and assesses what you need it to. This framework will help to assure quality material, but it is not a guarantee. While these systems are good, they are not infallible and your trained eye needs to be the final filter before using the output.

Exploring the Sample Prompt

This prompt demonstrates a clear and purposeful approach to generating an effective assessment, aligning well with the steps in the flowchart. The opening of the prompt establishes essential context by stating, "*I just finished teaching my 6th-grade social studies class a unit on European countries and capitals.*" By doing so, it ensures the AI understands the focus of the test, the grade level of the students, and the specific content to be assessed. This context allows the AI to tailor its output to match the learning objectives and grade-appropriate expectations.

Next, the prompt clearly defines the skills and knowledge to be evaluated: "*The test should evaluate students' ability to match countries to their capitals and identify countries based on their capitals.*" This specificity removes ambiguity and directs the AI to focus on generating questions that align precisely with the learning outcomes. By emphasizing these dual objectives—recognizing capitals and connecting them to countries—the educator ensures that the assessment comprehensively addresses the key components of the unit.

The structure of the assessment is also well-defined, with the request for "*20 questions, evenly split between multiple-choice and fill-in-the-blank formats, appropriate for 6th graders.*" This directive not only ensures variety in question types but also indicates the importance of making the assessment accessible to the age group. The balance between formats provides a mix of recognition and recall tasks, engaging students with different strengths and reinforcing their understanding through diverse approaches.

Finally, the prompt concludes with a practical and actionable element: "*Please include an answer key and ensure the questions align with the unit content.*" This reinforces the need for accuracy and alignment, ensuring that the generated output is directly usable without requiring significant edits. By requesting an answer key, the educator also emphasizes the importance of efficiency and readiness for immediate application.

This example highlights the effectiveness of breaking down assessment design into clear, actionable steps. Each component of the prompt contributes to producing a targeted, age-appropriate, and aligned resource that meets both instructional and evaluative needs. Educators can use this approach as a blueprint for crafting their prompts, ensuring that the AI delivers tailored outputs that enhance teaching and learning outcomes.

Adaptive Learning Flowchart

The Adaptive Learning Flowchart formula is best suited for queries that involve the analysis of student performance data and the personalization of learning pathways. This includes scenarios where educators need to receive AI-generated recommendations on how to adapt instruction to meet individual student needs, ultimately enhancing the learning experience for students.

This formula involves defining the educational objective, identifying the target audience, specifying the learning objectives, and determining the educational context. The prompt should be crafted to create a clear and specific prompt that instructs the AI algorithm on how to adapt instruction to meet individual student needs. This formula is designed to provide educators with personalized recommendations based on the analysis of student performance data, ultimately enhancing the learning experience for students. By applying the Adaptive Learning Pathways formula, a weaker prompt can be refined to include specific learning objectives, the intended target audience, and the educational context. This transformation ensures that the AI-generated recommendations are tailored to meet the specific educational needs and standards, ultimately enhancing its effectiveness and relevance.

Educators can used a variety of data to personalize learning pathways in the Adaptive Learning Pathways formula include various metrics such as students' assessment scores, learning progress, engagement levels, and responses to formative assessments. These data points are analyzed by AI algorithms to identify students' strengths, weaknesses, and learning preferences, allowing for the generation of personalized recommendations on how to adapt instruction to meet individual student needs. By leveraging this data, educators can gain insights into each student's unique learning profile and receive tailored guidance on how to optimize their learning experiences. This data-driven approach ensures that the learning pathways are adapted to meet the individual needs of students, ultimately enhancing the learning experience

The formula communicates with the AI by providing clear and specific instructions for the AI algorithm to follow. It guides the AI in understanding the educational objective, the target audience, the learning objectives, and the educational context, enabling the AI to generate personalized recommendations. By inputting a well-crafted prompt into the AI system, the AI uses this information to perform the analysis and generate personalized recommendations, ensuring that the learning pathways are adapted to meet the

Adaptive Learning Flowchart

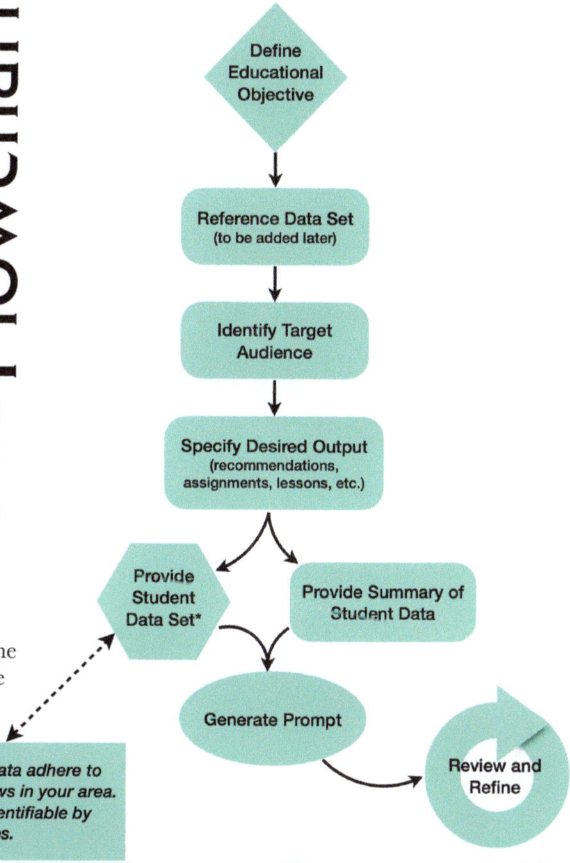

Scan to access the online image of the Adaptive Learning Flowchart

If using actual student data adhere to appropriate data privacy laws in your area. Make sure data is not identifiable by outside sources.

I am working with a group of high school sophomores early language learners. Their native language is German, and they are currently reading at a Lexile range of L800-L900 in English. I would like to adapt the attached reading material to suit their needs. The adaptation should simplify the text to align with their reading level in English while preserving the main ideas and cultural context. Additionally, include comprehension questions designed to assess understanding at their current proficiency level and a glossary of key terms and difficult words translated into German. Please ensure the revised material is accessible, engaging, and supportive of language acquisition for this group of students.

individual needs of students and enhance the learning experience.

The Adaptive Learning Pathways formula determines which learning objectives to personalize for individual students by analyzing student performance data and providing personalized recommendations. The prompt clearly defines the educational objective, which is to create adaptive learning pathways using AI, emphasizing personalizing learning pathways based on student performance data. It identifies the target audience as educators who will use the AI-generated recommendations to adapt instruction for individual students. The prompt instructs the AI algorithm to analyze student performance data and generate recommendations for educators to adapt instruction based on individual student needs. The AI uses this prompt to perform the analysis and generate personalized recommendations for educators, ensuring that the learning pathways are adapted to meet the individual needs of students and enhance the learning experience.

Using the Adaptive Learning Flowchart

Using the Adaptive Learning Flowchart is like having a roadmap for tailoring educational content to meet the diverse needs of your students. This tool breaks down the process into manageable steps, ensuring that the materials you create are accessible, engaging, and aligned with specific learning objectives. Whether you're adapting content for early language learners, advanced students, or those with unique learning needs, the flowchart provides a structured approach to guide your decisions.

The process begins by identifying your learners' specific needs. This includes understanding their current skill level, learning preferences, and any barriers they may face. For example, if you're working with students who are learning English as a second language, the flowchart encourages you to consider their proficiency in reading, writing, and comprehension. Taking this step ensures that the content you create meets your students where they are and provides a solid foundation for growth.

Defining the learning objectives is the next step in the process. This ensures that the adapted content aligns with your educational goals. Rather than creating general materials, the flowchart prompts you to focus on specific outcomes. If you aim to improve students' reading comprehension, for instance, it encourages you to incorporate key vocabulary terms, scaffolding techniques, or comprehension questions that target this skill. Having clear objectives keeps your efforts purposeful and provides a clear path for assessing effectiveness.

Adapting the content follows naturally, with a focus on simplifying or reformatting the material to make it more accessible while retaining its educational value. You might modify a complex article into shorter, more digestible sections or add visual aids for students who learn better through images. The flowchart emphasizes maintaining the integrity of the original material while transforming it to better suit your learners' needs.

The process concludes with evaluation and iteration. Once the content is adapted, it's essential to review it to ensure it aligns with your objectives and is appropriate for your students. Testing the materials in your classroom and gathering feedback can provide valuable insights. For instance, you might discover that the glossary you included was especially helpful or that the comprehension questions were too challenging. This feedback allows you to refine the materials, making them even more effective for future lessons.

The Adaptive Learning Flowchart is a practical tool for designing customized content that meets the unique needs of your students. It provides a structured yet flexible approach, helping educators create meaningful learning experiences without losing sight of their educational goals. By thoughtfully adapting materials, you can ensure that every student has the opportunity to succeed.

Exploring the Sample Prompt

The goal of this prompt is clear and thoughtfully articulated, making it an excellent example of how to effectively align educational objectives with student needs. The prompt begins by defining its purpose: "*I am working with a group of high school sophomores early language learners. Their native language is German, and they are currently reading at a Lexile range of L800–L900 in English.*" This introduction provides essential context, ensuring that the task is targeted and actionable. By specifying the reading level and linguistic background, the prompt sets a foundation for creating materials that directly support the students' language acquisition while meeting their academic requirements.

One of the prompt's strengths is its clear identification of the target audience. By explicitly stating that the students are German-speaking sophomores reading at a Lexile range of L800–L900, it equips the AI with detailed information about the learners' proficiency and needs. This clarity ensures that the adapted material will match their cognitive and linguistic abilities, creating an accessible and engaging learning experience. For example, if the AI understands that the students are still building their English vocabulary, it can avoid overly complex sentence structures while maintaining the text's main ideas.

The requested outputs in the prompt are another standout feature. It specifies, "*The adaptation should simplify the text to align with their reading level in English while preserving the main ideas and cultural context. Additionally, include comprehension questions designed to assess understanding at their current proficiency level and a glossary of key terms and difficult words translated into German.*" This level of detail gives the AI clear instructions, leaving no room for ambiguity. By breaking the task into actionable components—simplified text, comprehension questions, and a bilingual glossary—the prompt ensures that the output will be both comprehensive and practical.

The balance between simplification and preservation of context is also well-addressed. The prompt emphasizes that the adapted material should maintain the cultural and thematic integrity of the original text while making it accessible to the students. For instance, instead of merely translating the text into simpler English, the AI is encouraged to retain its essence, ensuring that the students gain a meaningful understanding of the content. This approach reflects an understanding of the importance of cultural sensitivity in language learning, as it helps students connect with the material in a way that goes beyond surface-level comprehension.

Finally, the prompt implicitly encourages an iterative review process. While refinement is not explicitly mentioned, the inclusion of deliverables such as comprehension questions and a glossary allows educators to evaluate the AI's response and make adjustments as needed. By asking for multiple types of outputs, the prompt supports a holistic adaptation process that can be further refined to meet classroom goals effectively.

This well-crafted prompt demonstrates how thoughtful design can transform AI into a powerful tool for education. It combines specificity, actionable goals, and cultural awareness to create materials that are not only tailored to the students' needs but also promote deeper engagement and understanding. Through its structured approach, it serves as a practical guide for educators looking to adapt resources for diverse learning environments.

PBL Experience Flowchart

Introducing the PBL Experiences Flowchart offers educators a clear, step-by-step guide to designing effective and engaging Project-Based Learning (PBL) lessons. Project-Based Learning empowers students to explore real-world problems through hands-on activities and inquiry-based methods, fostering critical thinking, collaboration, and creativity. However, creating PBL experiences that

PBL Experience Flowchart

Scan to access the online image of the PBL Experience Flowchart

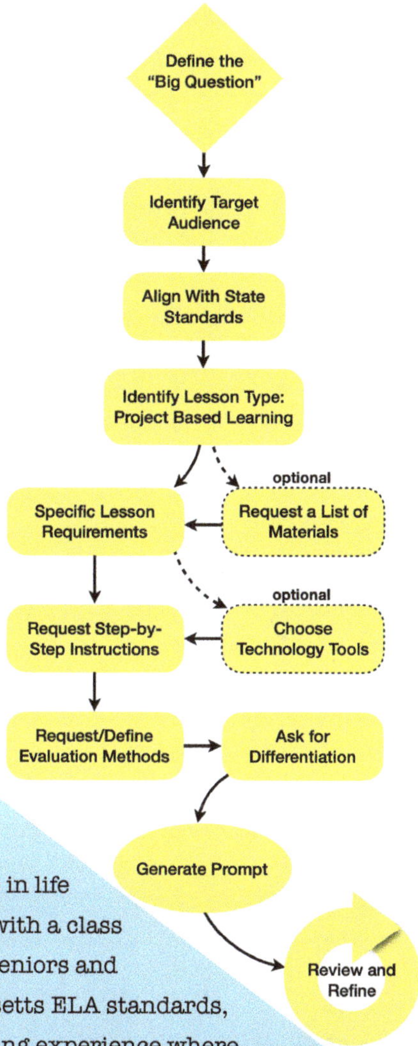

To what extent does success in high school determine success or failure in life after high school? Working with a class of honors-level high school seniors and aligning with the Massachusetts ELA standards, create a project-based learning experience where students are asked to complete action research and report their findings in some way. Provide a list of materials that students will need. Additionally, create step-by-step instructions for students to follow, which include benchmarks for periodic check-ins during the project. Include a rubric for evaluating the project. Provide differentiation for learners who may struggle with social interactions.

are meaningful and aligned with curriculum standards can be complex. This flowchart simplifies the process, providing a roadmap to ensure every aspect of the lesson, from objectives to assessments, is thoughtfully planned and executed.

The flowchart begins by emphasizing the importance of identifying the lesson type as PBL and specifying a topic. This foundation ensures the project has a focused direction and sets the stage for deeper exploration. For example, a topic like The Water Cycle provides a concrete starting point for students to investigate processes such as evaporation and precipitation. By clearly stating the learning objectives and aligning them with appropriate standards, such as the Next Generation Science Standards (NGSS), educators can ensure the project meets academic goals while promoting hands-on learning.

From there, the flowchart guides educators to include essential components such as step-by-step instructions, required materials, and detailed project descriptions. This ensures clarity and organization, both for teachers implementing the lesson and for students engaging with it. The inclusion of optional elements like differentiation models, technology integration, and extension activities enriches the learning experience, catering to diverse learners and making the project adaptable to varying classroom contexts.

Designed for both novice and experienced educators, the PBL Experiences Flowchart bridges the gap between theory and practice, providing a structured approach to crafting lessons that are engaging, relevant, and educationally robust. Whether you're teaching middle school science or adapting the framework for other disciplines, this tool equips you with the strategies needed to make project-based learning a powerful component of your classroom.

Using the PBL Experience Flowchart

At its core, the process starts with identifying the "Big Question," which acts as the driving force for your project or lesson. A strong Big Question is more than just a topic—it's an inquiry that provokes thought, challenges assumptions, and connects to real-world issues. When crafting this question, consider what will inspire curiosity in your students and lead to meaningful exploration. A well-constructed question encourages critical thinking, collaboration, and problem-solving, setting the tone for the entire learning experience.

Understanding your target audience is equally important. Who your students are—their grade level, abilities, and interests—will shape every decision you make. By keeping their needs at the forefront, you can ensure the lesson is accessible and engaging. For instance, if your students excel at hands-on activities but struggle with traditional lectures, this is where you begin adapting the project to fit their strengths. Recognizing your audience also helps you identify challenges they might face, whether it's a need for additional scaffolding, technology integration, or differentiated support.

Aligning the lesson with state or national standards is not just about meeting requirements—it's about grounding your lesson in skills and competencies that matter. This ensures the project is purposeful and provides opportunities for students to develop academically while addressing broader learning goals. Instead of viewing standards as a limitation, use them as a framework to design a lesson that integrates creativity and inquiry with measurable learning outcomes.

The type of lesson you design plays a significant role in its success. In this flowchart, project-based learning is highlighted, which prioritizes active engagement and real-world application. However, within this framework, there's room for flexibility—your lesson could lean toward research, collaboration, artistic creation, or even problem-solving simulations. When deciding on your lesson type, reflect on how your students learn best and what outcomes you want to achieve.

The flowchart then guides you into the practical aspects of lesson design. At this stage, it's time to outline the specific requirements. What materials will students need to succeed? Will you require physical tools, digital resources, or a mix of both? Thinking through this carefully ensures that students are equipped to engage fully in the process. Simultaneously, consider the role of technology. Tools like collaborative platforms, research databases, or creative software can amplify the lesson's impact and help students express their ideas in new ways.

Clear, detailed instructions are essential for guiding students through the project. Providing step-by-step guidance ensures that students know what is expected at every stage, reducing confusion and fostering independence. It's helpful to incorporate benchmarks or periodic check-ins—these milestones allow students to reflect on their progress, adjust their work if needed, and receive feedback before the final product is due. This structure keeps students motivated and ensures the lesson progresses at a manageable pace.

Evaluation is another key piece of the flowchart's process. By creating a rubric or another clear method of assessment, you give students a transparent understanding of how their work will be judged. A well-designed rubric not only outlines expectations but also serves as a learning tool, helping students understand how to improve their work and meet the objectives. Evaluation methods should also be aligned with the project's goals—whether that means focusing on creativity, collaboration, critical thinking, or another skill.

An often-overlooked but critical component of lesson design is differentiation. The flowchart emphasizes the need to consider learners who may face unique challenges, such as those struggling with social interactions. Differentiation could mean providing alternative tasks, allowing for varied methods of presenting work, or offering additional support where needed. It's an opportunity to make the project inclusive, ensuring that every student feels empowered to succeed.

The final stages of the flowchart tie everything together. Once you've generated the prompt, it's important to review and refine your plan. This isn't just about catching mistakes—it's about taking a step back and asking whether your lesson truly meets the needs of your students. Are the goals clear? Is the project engaging? Does it leave room for creativity while maintaining structure? This reflective step allows you to refine the lesson into something truly impactful.

By following this flowchart thoughtfully, you're not just creating a lesson—you're crafting a meaningful learning experience that challenges, supports, and inspires your students. Each step in the process adds depth and clarity, ensuring that the lesson is purposeful and effective. As you work through it, remember that flexibility is key. Use the flowchart as a guide, but let your creativity and understanding of your students shape the final product.

Exploring the Sample Prompt

When we compare the sample prompt to the framework, we see a strong alignment across its key components. The framework emphasizes the importance of defining a "Big Question" to drive inquiry and focus the learning process. In this case, the prompt centers on a thought-provoking essential question: "*To what extent does success in high school determine success or failure in life after high school?*" This question is open-ended, rooted in real-world relevance, and invites critical thinking, which is integral to Project-Based Learning (PBL).

The second step in the framework involves identifying the target audience. The prompt specifies "*honors-level high school seniors*" as the audience, making it clear who the project is designed for. By tailoring the activity to this group, the instructor ensures the complexity and expectations are appropriate for the students' capabilities while also challenging them to explore abstract concepts such as "success" and "failure."

The alignment with state standards, as noted in the framework, is also explicitly addressed in the prompt: "*aligning with the Massachusetts ELA standards.*" By incorporating these standards, the project maintains academic rigor and provides clear learning objectives that are measurable and transferable. This integration not only meets educational requirements but also ensures the activity builds essential skills like research, analysis, and communication.

When the framework moves into identifying the lesson type, the prompt's directive to "*create a project-based learning experience where students are asked to complete action research and report their findings*" demonstrates a commitment to authentic, student-driven inquiry. Action research emphasizes real-world problem-solving and active engagement, making it an ideal match for PBL. This approach empowers students to not only learn about the topic but also to engage in data collection, analysis, and meaningful reflection, all of which promote deeper learning.

The framework's next steps—defining lesson requirements, listing materials, and providing instructions—are well-supported in the prompt. For example, the prompt explicitly calls for "*a list of materials that students will need*" and "*step-by-step instructions for students to follow, which include benchmarks for periodic check-ins.*" These benchmarks create opportunities for scaffolding and formative assessment, ensuring that students remain on track and receive guidance as needed. Providing a clear structure for the project allows students to navigate the process effectively, which is critical in PBL.

Moreover, the framework suggests including evaluation methods, a point explicitly addressed in the prompt: "*Include a rubric for evaluating the project.*" A rubric provides transparency in grading and helps students understand expectations, which aligns with the framework's emphasis on evaluation. Additionally, the inclusion of differentiation strategies, as noted in the prompt—"*Provide differentiation for learners who may struggle with social interactions*"—highlights a commitment to Universal Design for Learning (UDL) principles. By accommodating diverse learners, the project ensures that all students can participate meaningfully, fostering equity and inclusion.

The framework's iterative nature is captured in the prompt's emphasis on review and refinement, particularly through the benchmarks and rubric. These elements encourage both students and teachers to reflect on progress and make adjustments, which are hallmarks of effective PBL. The prompt closely adheres to the PBL framework by defining a compelling "Big Question," identifying the target audience, aligning with standards, incorporating action research, and ensuring a clear, scaffolded process for students. Its emphasis on differentiation and evaluation further supports the principles of effective PBL, making it a well-rounded approach to fostering critical thinking, collaboration, and real-world application.

Case Study:

Choosing the Right Framework for Educational Prompts

I. **Scenario:** Ms. Taylor, a high school social studies teacher, is preparing a unit on global trade for her sophomore students. Her objectives are threefold: to assess students' understanding of global trade concepts, align the unit with Massachusetts state standards, and create an engaging, hands-on project that encourages critical thinking and collaboration. Ms. Taylor knows her students perform best when assessments and projects are clear, structured, and tailored to their needs, so she decides to use the available frameworks to guide her planning process.

II. **Identifying the Primary Goal:** Ms. Taylor begins by identifying her key objectives for the unit. She needs to evaluate her students' knowledge of global trade, engage them with real-world applications, and ensure her materials meet state educational standards. This leads her to select three frameworks: the *Assessment Item Generation Flowchart* for creating a summative test, the *PBL Experiences Flowchart* for designing an interactive project, and the *Curricular Alignment Flowchart* to verify compliance with standards.

III. **Creating an Assessment Using the Assessment Item Generation Flowchart:** To measure her students' understanding, Ms. Taylor uses the *Assessment Item Generation Flowchart*. This framework helps her design a summative assessment that evaluates key concepts and critical thinking skills.

A. *Prompt Example:*

> Develop a 20-question assessment for a high school social studies class covering global trade. The test should include 10 multiple-choice questions assessing key trade concepts and 10 short-answer questions analyzing trade's economic and cultural impacts. Provide an answer key and ensure alignment with Massachusetts social studies standards.

B. *Why This Framework Was Chosen:* Ms. Taylor wants a robust test that captures both factual knowledge and analytical skills. The flowchart ensures her assessment is well-rounded, appropriately challenging, and aligned with the curriculum.

IV. **Engaging Students Through Project-Based Learning Using the PBL Experiences Flowchart:** To deepen her students' engagement, Ms. Taylor plans a project where they can explore global trade networks. She uses the *PBL Experiences Flowchart* to ensure the project is structured, accessible, and aligned with her educational goals.

V. *Prompt Example:*

> Design a project-based learning experience for high school students to explore global trade. Students will research a trade route (e.g., the Silk Road or modern shipping lanes) and create a visual presentation analyzing the route's economic and cultural impacts. Include step-by-step instructions, benchmarks for progress, and a rubric for assessment. Ensure alignment with state standards and provide differentiation for students needing additional support.

A. *Why This Framework Was Chosen:* The project offers an opportunity for active learning and collaboration, key elements of engagement. The *PBL Experiences Flowchart* helps Ms. Taylor define clear steps, provide support for diverse learners, and align the project with state standards.

VI. **Ensuring Alignment Using the Curricular Alignment Flowchart:** Finally, Ms. Taylor reviews her materials using the *Curricular Alignment Flowchart.* This framework ensures both her assessment and project meet Massachusetts social studies standards and identifies any areas needing adjustment.

A. *Prompt Example:*

> Analyze Massachusetts social studies standards for high school global trade content and compare them with the attached unit plan. Highlight any gaps in alignment and suggest adjustments to ensure both the assessment and project meet state requirements.

B. *Why This Framework Was Chosen:* By confirming alignment with state standards, Ms. Taylor ensures her unit is not only engaging but also compliant with educational requirements, reinforcing the relevance and rigor of her teaching.

VII. **Conclusion:** Ms. Taylor demonstrates how selecting the right frameworks for specific teaching objectives can create a cohesive and effective unit. The Assessment Item Generation Flowchart helps her design a strong evaluation tool, the PBL Experiences Flowchart supports her in creating an engaging hands-on project, and the Curricular Alignment Flowchart ensures her materials meet state standards. By combining these frameworks, Ms. Taylor delivers a unit that is both rigorous and engaging, helping her students succeed in understanding global trade concepts.

Choosing Between the Frameworks

Selecting the right framework for your educational goals can transform your teaching experience, ensuring your efforts are purposeful and impactful. Each framework has unique strengths and is tailored to address specific tasks, whether you're creating assessments, designing project-based learning experiences, or ensuring curriculum alignment. To choose the best framework—or combine several effectively—it's essential to align your decision with your objectives, your students' needs, and the broader learning context.

Start by clearly defining your goal. For instance, if you aim to evaluate student knowledge with precision and variety, the *Assessment Item Generation Flowchart* is your go-to. It guides you through crafting standards-aligned quizzes or tests with diverse question types and clear objectives. On the other hand, if you want to engage students in hands-on, collaborative learning, the *PBL Experiences Flowchart* offers a roadmap for creating projects that emphasize active participation and real-world application. Meanwhile, if your priority is ensuring lessons meet state or national standards, the *Curricular Alignment Flowchart* can help you pinpoint areas of strength and gaps in your content. By identifying your primary objective, you can select the framework that aligns most closely with your needs.

Consider the learning context and your students' specific requirements. For diverse classrooms where students have varying strengths and challenges, combining frameworks can be highly effective. For example, you might use the *Curricular Alignment Flowchart* to ensure your lessons meet required benchmarks while incorporating the *PBL Experiences Flowchart* to foster adaptability and engagement. This approach ensures that your lesson planning is both rigorous and inclusive, addressing academic goals while supporting students who may need differentiated instruction.

Differentiation is another key factor in choosing a framework. The *PBL Experiences Flowchart*, for instance, includes built-in strategies for tailoring projects to meet individual needs, making it ideal for supporting students with alternative learning styles. If your classroom includes learners who struggle with traditional methods, frameworks that emphasize flexibility and engagement can be invaluable. Think about how each framework supports your ability to provide scaffolding, individualized pacing, or alternative approaches to content delivery.

Integrating frameworks can also create a cohesive teaching plan. Let's say you're introducing a new unit. Start with the *Curricular Alignment Flowchart* to ensure the unit aligns with state standards. Then, use the *PBL Experiences Flowchart* to design an interactive project that encourages students to explore the topic deeply. Finally, rely on the *Assessment Item Generation Flowchart* to develop a summative assessment that measures both their understanding and application of the material. Combining frameworks ensures that every aspect of teaching—engagement, alignment, and evaluation—is addressed in a unified way.

When using these frameworks, balance rigor with engagement. Effective teaching requires meeting high academic standards while keeping students motivated and involved. The frameworks are tools, not rigid rules, so adapt them to fit your classroom dynamics and priorities. They're most effective when used flexibly to support practical teaching scenarios, such as creating clear instructions, rubrics, or progress benchmarks that guide students through the learning process.

Try it yourself: Choosing the Right Framework

- **Clarify Your Goal:** Define whether you're creating assessments, designing a project, or aligning curriculum.
- **Choose the Framework:** Match your goal to the most relevant framework, such as *Assessment Item Generation* for quizzes or *PBL Experience* for projects.
- **Follow the Steps:** Apply the framework step-by-step, tailoring it to your needs.
- **Refine and Reflect:** Adjust your work based on the framework's guidance and evaluate its effectiveness for future use.

Chapter 7: Iterative Refinement

Iterative refinement in prompt engineering is like the careful work of a sculptor, shaping raw material into something precise and purposeful. Your first prompt is a block of stone or wood—full of potential but needing direction. Each revision is a careful cut or smoothing stroke, bringing the form closer to your intended vision. The process involves guiding the AI with increasing clarity and specificity, transforming its responses from broad and undefined to sharp and actionable.

In the beginning, the work focuses on rough shaping. A broad prompt might create a response that is functional but lacks focus, much like the early stages of carving reveal only the basic outline of a figure. For example, starting with "*Explain the role of accommodations in special education*" might produce a general overview, which is useful as a foundation. But when your goal requires detail and context, the shaping continues. Adjusting the prompt to "*Provide a list of five accommodations for high school students with ADHD to enhance executive functioning, including implementation examples*" gives the AI the structure it needs to deliver something more aligned with your needs.

As you refine further, the process transitions to adding detail and polish. This is the stage where small adjustments make a significant difference. Including an intended audience, the purpose of the output, or even the format—such as a slide deck or quiz—helps the AI craft responses that are specific and immediately usable. These refinements are like smoothing rough edges or defining intricate details, ensuring the response isn't just accurate but tailored to your teaching goals.

There is also a stage for stepping back and examining the work. Sculptors often pause to view their progress from different angles, adjusting where necessary to maintain balance and harmony. In the same way, reviewing the AI's responses allows you to assess their depth, tone, and alignment with your objectives. If the output lacks nuance or misses the mark, further adjustments—such as rephrasing questions, adding background information, or narrowing the focus—can bring it closer to your desired outcome.

In its final stages, the refinement process turns your initial work into a polished and effective tool. Each adjustment hones the prompt's ability to communicate your needs to the AI, ensuring that the responses are not only clear and relevant but also aligned with your classroom objectives. Like a completed sculpture, a well-refined prompt stands as a testament to the skill and care invested in its creation.

The art of prompt refinement is an ongoing process, one that grows with practice and attention to detail. Each prompt you

craft and refine is a step toward mastering this interaction, making AI a seamless and valuable partner in achieving your educational goals. Through this process, you aren't just producing better outputs—you're developing a skill that transforms how you approach teaching and learning in a world increasingly influenced by technology.

Case Study:

Discovering the Power of Iterative Refinement

I. **Scenario:** Samantha, a high school business teacher, wanted to integrate AI into her curriculum by creating a project for her students on business ethics. Excited to explore new possibilities, she crafted her first prompt: "*Explain business ethics.*" Expecting a detailed response that would jumpstart her lesson planning, Samantha was frustrated when the output was too broad, listing generic definitions and concepts without depth or actionable insights.

II. **Initial Challenges**

   A. *Generalized Responses*: The AI provided a basic overview of business ethics but failed to offer specific examples or practical applications relevant to her class.

   B. *Misalignment with Objectives*: Samantha needed content tailored to high school students, but the output felt more appropriate for a general audience.

C. *Lack of Depth*: The information lacked the nuance and engagement necessary to spark meaningful discussions in class.

III. **First Refinement:** Samantha decided to add specificity to her prompt. She revised it to: "*Explain three examples of ethical dilemmas businesses face and how they can address them.*" While the AI's response improved, offering real-world examples like workplace discrimination and environmental responsibility, the tone was still too formal and lacked student engagement.

A. *Improved Focus*: The examples provided were relevant but still generic, leaving Samantha to bridge the gap between the content and her students' needs.

B. *Tone Issues*: The AI's language felt academic and dry, which wasn't suitable for high schoolers.

C. *Opportunity for Context*: The prompt lacked details about her audience and intended format, which limited the AI's ability to deliver actionable teaching materials.

IV. **Second Refinement:** Determined to improve the results, Samantha tried again. This time, she added more context and specified her desired outcome: "*Create a slide deck for high school students on three ethical dilemmas businesses face, using relatable examples and discussion questions to engage students.*"

A. *Tailored Output*: The AI's response aligned better with her objectives, presenting relatable case studies like social media transparency and fair labor practices.

B. *Engaging Content*: The inclusion of discussion questions helped bridge the gap between abstract concepts and classroom dialogue.

C. *Usable Format*: By specifying a slide deck, Samantha received a structured response that was ready to integrate into her lesson.

V. **Final Results:** Samantha reviewed the AI's output and made small tweaks to improve tone and clarity, such as asking the AI to "use age-appropriate language" and "include visuals for each example." The final result was a polished, engaging, and student-centered resource that aligned perfectly with her teaching goals.

VI. **Key Takeaways**

A. *Start Broad, Then Refine*: Samantha's journey demonstrates the importance of starting with a general prompt and iterating to add focus, context, and format.

B. *Specify Tone and Audience*: Including the target audience and desired style helps the AI tailor responses effectively.

C. *Leverage Iteration for Engagement*: Adjustments to the prompt encouraged the AI to provide actionable and engaging content, enhancing classroom discussions.

Through the iterative refinement process, Samantha transformed frustration into a learning opportunity, mastering a skill that not only improved her AI interactions but also enriched her teaching practices.

Evaluating AI Responses

Refining AI prompts is an iterative process, much like sculpting a detailed work of art. Each evaluation of an AI-generated response is an opportunity to identify areas that need shaping, ensuring the final product meets your teaching objectives. Start by examining the accuracy of the content. Just as a sculptor ensures each cut aligns with their vision, educators must verify the AI's information against credible sources. For example, if an AI-generated response includes historical data, cross-check it with trusted references to confirm its reliability. This not only safeguards the lesson's integrity but also models critical evaluation for students.

Take a close look at how ***accurate*** the AI's response aligns with what you originally asked for and whether it truly adds value to your teaching goals. This step helps keep the content focused and useful. If the response feels scattered or too general, it's worth revisiting your prompt to make it more precise. For example, if you ask something broad like:

> Explain the causes of climate change.

you might get a response that covers too much without getting into the details you really need. Instead, try refining it to something like:

> Describe three human activities contributing to climate change and their impacts on biodiversity for a high school science class.

This narrows the scope and makes it clear who the audience is and what kind of content you're looking for. Refining prompts like this is about guiding the AI to focus on what matters most, saving you time and making the results more practical for your classroom. It's an ongoing back-and-forth process, almost like a conversation, where each tweak brings you closer to responses that really hit the mark.

When considering the iterative process of refining prompts, ***depth*** plays a critical role in shaping outcomes that genuinely foster student engagement and learning. Take, for example, the initial prompt:

> Explain the impact of urbanization on local wildlife.

While clear, this phrasing might generate a surface-level response that lacks depth or practical application. By refining it to:

> Create a chart and write a short explanation of how urbanization impacts three specific species in the New England area, focusing on changes to their habitats and food sources.

the prompt transforms into a task that not only demands critical thinking but also requires students to synthesize information and apply it to real-world contexts. This iterative refinement aligns the task more closely with higher-order cognitive skills, pushing students beyond basic understanding. By iterating prompts in this way, you're not just shaping AI responses; you're crafting experiences that guide students toward richer, more impactful learning opportunities.

Through trial and adjustment, educators can ensure their prompts align with specific goals, fostering critical analysis and creativity in their classrooms.

The iterative process of refining prompts isn't just about increasing clarity or specificity; it's also about enhancing ***tone and style*** to make tasks more engaging and relatable for students. For instance, starting with a prompt like:

> Analyze three leadership styles, providing examples of how each affects team dynamics in a corporate environment.

delivers straightforward information but may lack the energy needed to captivate students. By refining this to:

> Craft three engaging, real-world scenarios that showcase different leadership styles in action. For each scenario, include a detailed analysis of how the chosen style influences team morale, communication, and productivity. Add three thought-provoking questions for peers to reflect on which leadership style they find most effective and why.

the tone becomes more dynamic and interactive. Words like "*engaging*," "*real-world*," and "*thought-provoking*" invite students to actively connect with the material, rather than passively absorb it. This process demonstrates how adjusting tone and style can transform a dry, instructional prompt into an immersive activity that encourages critical thinking and deeper engagement. By iterating in this way, educators can create prompts that not only align with learning objectives but also resonate with students' curiosity and drive to participate.

Once you've evaluated these elements, consider areas for ***further refinement***. Identify gaps or missed opportunities in the AI's response and adjust the prompt accordingly. For example, if a response about executive functioning in special education lacks actionable strategies, refine the prompt to:

> **Provide three research-backed strategies for improving executive functioning in high school students with ADHD, including classroom implementation examples.**

This iterative process guides the AI toward generating precise, actionable content.

By evaluating and refining AI responses, educators can unlock the technology's potential as a dynamic classroom ally. This process not only enhances the utility of AI but also ensures it complements teaching strategies, fosters critical thinking, and supports student engagement. Just as sculptors work patiently to reveal the beauty within the stone, educators can refine prompts to reveal the most valuable insights from AI.

Refining AI-generated responses is a process that requires careful evaluation and intentional adjustments. Whether you're using AI to generate lesson plans, student resources, or assessments, the goal is to ensure the content meets high standards of accuracy, relevance, and engagement. By focusing on these key criteria—

accuracy, relevance, depth, tone, and opportunities for improvement—you can transform AI into a powerful tool that supports both teaching and learning. These considerations ensure the outputs align with educational goals while fostering critical thinking and meaningful interaction in your classroom.

1. **Verify the Accuracy** of AI-generated responses by cross-checking information against credible sources and prior knowledge to ensure the reliability of data provided to students.
2. **Assess the Relevance** of AI responses to ensure they directly address the query and align with educational goals.
3. **Depth and Complexity** of AI responses should be evaluated using Bloom's Taxonomy as a benchmark to ensure they promote higher-order thinking skills.
4. **Tone and Style** in AI responses should be checked to ensure they are consistent, engaging, personalized, and enhance the learning experience.
5. **Opportunities for Refining** AI prompts to elicit more precise and enriching responses should be identified, enhancing the AI's contribution to learning.

Mastering the iterative process of refining AI responses goes beyond simply verifying facts. It's about ensuring that the outputs are not only accurate but also aligned with your teaching objectives, engaging for your students, and capable of fostering higher-order thinking skills. By regularly assessing and adjusting prompts, you unlock the true potential of AI as an educational partner, crafting experiences that are tailored, impactful, and transformative. This approach empowers educators to use AI not just as a tool, but as a collaborator in delivering high-quality, dynamic learning opportunities.

Iterative refinement is about continuous improvement. By evaluating AI responses for accuracy, relevance, depth, and tone, and then revising your prompts accordingly, you can achieve more precise and actionable outcomes. This process requires time and patience, but the results—engaging, tailored, and effective AI-generated content—are worth the effort. Whether you're creating differentiated assignments, designing quizzes, or developing project-based learning experiences, iterative refinement ensures that AI becomes a valuable and reliable partner in your teaching journey.

Try it yourself: Refining Prompts for Depth and Clarity

- **Begin with a Broad Prompt:** Start with a general question to use as your baseline to assess the AI's initial output.
- **Evaluate the AI's Response:** Carefully review the AI's response for:
  - *Accuracy*: Are the facts correct?
  - *Relevance*: Does it address the question?
  - *Depth*: Is the explanation thorough or too surface-level?
  - *Tone*: Does the response fit the intended audience and purpose?
- **Refine Your Prompt:** Rewrite the original prompt to add clarity, depth, and context.
- **Run the Refined Prompt:** Submit your improved prompt to the AI and compare the new output to the original. Look for specificity, alignment with your goals, and improved tone.
- **Reflect and Iterate:** Reflect on the differences between the outputs. What worked better in the refined prompt? If needed, continue refining to further tailor the response to your teaching objectives.

Chapter 8: Creative Applications

Using AI creatively in the classroom has the potential to transform education by fostering deeper engagement, encouraging critical thinking, and sparking creativity. By understanding what AI is capable of, educators can develop specific and detailed prompts that guide AI to generate content tailored to their learning objectives. This approach not only creates a more interactive and personalized learning environment but also inspires students to think critically and creatively.

AI can inspire innovative assessment ideas by analyzing learning outcomes and generating diverse suggestions for evaluating student progress. For instance, inputting a goal like:

Assess understanding of renewable energy concepts.

might yield suggestions such as designing a quiz, creating a poster project, or organizing a group debate. These ideas can be refined iteratively to better align with classroom goals, saving time and introducing creative ways to evaluate learning.

Additionally, AI can provide personalized feedback, allowing students to apply their knowledge in a safe, guided environment. For example, students might submit their written work to an AI tool that offers constructive suggestions for improvement. This interactive approach encourages self-reflection and collaborative learning, making the classroom more dynamic and engaging.

Employing AI to generate analogies and examples that relate to students' interests—like TikTok videos or popular movies—can further enhance engagement. While these analogies may not be perfect, they serve as starting points for student groups to critique and build upon, enhancing their analytical and creative skills. Using AI-generated content as a springboard for deeper exploration exemplifies how prompt engineering can stimulate critical thinking and creativity in the classroom.

It's important to integrate AI in education in a manner that prioritizes safety, equity, and effectiveness. The Office of Educational Technology emphasizes developing policies and supports focused on the safe, fair, and effective use of AI-enabled educational technology. Effective prompt engineering aligns with these priorities by equipping educators to create tailored prompts that improve the quality of AI-generated content and contribute to a safe and supportive learning environment.

By incorporating prompt engineering into educational initiatives, schools can ensure that the use of AI in the classroom is not only safe and effective but also tailored to meet the specific needs of students and learning objectives. This approach empowers educators to harness the full potential of AI in education while maintaining a strong focus on safety, equity, and effectiveness.

Exploring Storytelling with AI

Storytelling has always been a cornerstone of education, and integrating AI into this process introduces a fresh, interactive dimension. AI tools can generate story prompts, draft narratives, or even simulate characters, helping students explore new creative avenues. For example, a teacher might prompt an AI to generate a *"short story featuring a character who must solve an ethical dilemma in a futuristic society."* The result serves as a starting point, with students critiquing and expanding the narrative to add their unique perspectives. By collaborating on these AI-generated foundations, students deepen their understanding of narrative structures, character development, and thematic expression.

AI's storytelling applications go beyond creative writing. Educators can use AI to craft dynamic historical or scientific narratives that adapt based on student input. For instance, in a history class, students might guide an AI-generated historical figure through key events, exploring the motivations and consequences of their decisions. This interactive format not only enhances engagement but also helps students internalize complex concepts by participating in the storytelling process.

Case Study:

Revitalizing Creativity in Science Education

I. **Scenario:** Ms. Patel, a high school science teacher, noticed her students were excelling at memorizing facts but struggled with creative problem-solving. She wanted them to move beyond rote learning and engage in deeper, more innovative thinking about scientific concepts. Determined to address this, she decided to integrate AI into her classroom to encourage creative exploration and critical thinking.

II. **Approach:** Ms. Patel began by using AI tools to develop prompts that would foster creativity and engagement. She crafted an initial broad prompt:

> Explain how renewable energy works.

The AI's response provided a standard explanation of wind, solar, and hydropower technologies. While accurate, the output lacked the depth and creativity she wanted her students to experience.

To refine the approach, she experimented with a more targeted prompt:

> Describe three innovative ways to integrate renewable energy into urban environments, considering economic and environmental factors.

The response was more aligned with her goal, offering ideas like solar-panel sidewalks, vertical wind turbines, and algae-based biofuel systems.

Inspired, Ms. Patel adapted this output into a classroom activity. Students were tasked with critiquing the AI's suggestions, brainstorming improvements, and creating their own innovative renewable energy solutions. They worked in groups, presenting ideas supported by AI-generated visuals and analyses.

III. **Outcome:** The shift in approach invigorated Ms. Patel's classroom. Students engaged deeply with the material, creatively applying scientific principles and debating the feasibility of their ideas. By integrating AI into her teaching strategy, Ms. Patel fostered a collaborative and imaginative learning environment.

IV. **Analysis:**

A. *Initial Challenges:* Ms. Patel identified that her students excelled at factual recall but lacked creative engagement. The initial AI prompt provided accurate information but did not inspire innovative thinking.

B. *Refinement Through Iteration:* By revising her prompt to include a specific context and challenge, Ms. Patel received more dynamic and actionable AI outputs that encouraged critical thinking.

C. *Student Engagement:* The refined prompts led to interactive group activities, where students not only critiqued the AI-generated ideas but also contributed their own, fostering collaboration and creativity.

D. *Practical Implementation:* Ms. Patel demonstrated how educators can effectively use AI to transition from static, fact-based instruction to engaging, problem-solving activities tailored to real-world applications.

V. **Takeaways:**

A. *Refining Prompts for Creativity:* Broad prompts can provide general information but often fail to inspire innovation.

Refining prompts with specific contexts and challenges encourages deeper thinking and creativity.

B. *Encouraging Collaboration:* AI-generated ideas can serve as a springboard for group activities, where students critique, expand, and improve on initial suggestions, fostering teamwork and engagement.

C. *Real-World Applications:* Tying prompts to practical challenges, like renewable energy solutions for urban settings, connects classroom learning to real-world issues, making lessons more relevant and impactful.

D. *Iterative Refinement Matters:* Experimenting with and refining prompts is essential for aligning AI outputs with teaching goals. Specificity, context, and purpose are key factors in achieving meaningful results.

E. *Empowering Students Through AI:* Using AI in a guided way empowers students to think critically, engage with technology, and explore creative solutions, preparing them for challenges beyond the classroom.

VI. **Conclusion:** This experience underscored the importance of iterative refinement in crafting AI prompts. By experimenting with specificity and relevance, Ms. Patel transformed her classroom into a space where creativity flourished alongside scientific inquiry. Her story illustrates how AI can serve as a catalyst for innovation when educators take a thoughtful and deliberate approach.

Data Analysis Through AI Prompts

AI can simplify data analysis, transforming raw information into actionable insights for both educators and students. Teachers can craft prompts that guide AI to analyze classroom data, such as:

> Identify trends in student performance on the latest assessments and suggest interventions for common misconceptions.

This use of AI enables data-driven decision-making, ensuring that instructional strategies are targeted and effective.

Students can also benefit from AI-assisted data analysis. For example, in a business class, a prompt like:

> Analyze sales data to identify market trends and propose a marketing strategy for a new product

can introduce real-world applications of statistical concepts. By guiding students through the process of interpreting AI-generated insights, educators encourage critical thinking and practical application, making learning more relevant and impactful.

Problem-Solving with AI Assistance

AI can be a valuable resource for breaking down complex problems into manageable steps. In subjects like math, science, and humanities, educators can craft prompts that encourage analytical thinking and creative solutions. For instance, a teacher might ask:

Develop a step-by-step plan for reducing water waste in urban areas, supported by data and real-world examples.

The AI's response provides a framework for students to critique, refine, and expand upon, fostering deeper understanding and engagement.

In the humanities, prompts might challenge students to explore alternative perspectives, such as:

Write a letter from the perspective of a historical figure defending their controversial decisions.

This exercise not only sharpens analytical skills but also helps students empathize with diverse viewpoints, enriching their learning experience.

AI's creative potential in education is vast, offering tools that enhance creativity, encourage critical thinking, and provide personalized learning experiences. By thoughtfully integrating AI into teaching practices, educators can transform traditional methods into dynamic, interactive, and inclusive learning environments. However, this requires a balance—leveraging AI's strengths while addressing ethical considerations like privacy and bias. With continued practice and refinement, educators can harness AI to inspire, engage, and prepare students for the challenges of a rapidly evolving world.

Try it yourself: Crafting Creative Prompts

Challenge Yourself: Use these five creative exercises to develop prompts tailored to your subject area:

- *Subject Integration:* Combine topics from two subjects. For example: environmental science and economics
- *Real-World Connection*: Link classroom content to current global issues. For instance: a marketing campaign and the global decline in bee populations
- *Character Perspective*: Dive into historical or literary viewpoints.
- *Creative Problem-Solving*: Encourage innovative exploration.
- *Cultural Exploration*: Investigate cultural perspectives.

These exercises sharpen your prompt-crafting skills while making lessons more engaging and meaningful for students.

Chapter 9:
Feedback and Analysis

Feedback is the bridge between mediocrity and mastery in AI prompt engineering. Think of it as a chef refining a recipe—the adjustments you make based on the feedback you receive can transform a decent prompt into an exceptional one. By assessing the output generated by AI and making necessary refinements, educators can craft prompts that better align with their objectives, fostering deeper engagement and more meaningful learning experiences. This chapter explores the role of feedback in refining AI prompts, offers strategies for building feedback communities, and provides practical exercises for honing this essential skill.

Feedback in AI prompt engineering comes in two primary forms: human feedback and updates to the dataset provided in the chat. Human feedback involves assessing the relevance, accuracy, and clarity of AI-generated responses, offering specific critiques to guide the AI closer to your intended outcome. Dataset updates incorporate new information or corrections to tailor responses to meet your needs better. Together, these approaches create a feedback loop that enhances the performance and reliability of AI systems, ensuring they deliver educationally relevant and impactful content.

Building a Community for Feedback

Feedback doesn't exist in a vacuum; it thrives in collaborative environments. Much like sharing lesson plans with colleagues, educators can benefit from a community of peers to evaluate and refine AI prompts. These colleagues should ideally bring diverse strengths, such as expertise in instructional design, familiarity with AI tools, and a knack for constructive criticism. Diversity in experience and perspective enriches the feedback process, offering varied insights that can improve both individual prompts and the broader application of AI in education.

When presenting prompts for feedback, clarity is key. A vague or overly specific prompt can lead to ineffective AI responses, making it challenging to extract meaningful or relevant content. Think of the initial prompt as the sketch for a painting: too vague, and it becomes a shapeless outline; too detailed, and it leaves no room for interpretation or creativity. For example, a prompt like:

> Explain the importance of art.

may result in a broad, generic answer that lacks depth or relevance. In contrast, if you refined the prompt to provide a specific direction while still leaving space for nuanced and engaging responses, you may end up with something like:

> Describe three ways the Impressionist movement influenced modern art techniques, with examples that could inspire a high school art project

This balance allows feedback providers to focus on refining the prompt's effectiveness rather than struggling to decipher its intent. By presenting clear and contextually relevant prompts, you set the stage for meaningful feedback and ensure that the iterative process builds on a solid foundation.

It's also essential to guard against biases in prompts and AI-generated responses. A good feedback community should identify potential issues, ensuring that prompts are inclusive and free from unintended assumptions. Encouraging reviewers to provide actionable, balanced critiques fosters a collaborative environment where feedback is seen as an opportunity for growth rather than criticism.

Analyzing AI Responses and Peer Feedback

Evaluating AI responses is not just about checking for correctness; it's about digging deeper to assess their accuracy, relevance, depth, and tone. For instance, if an AI-generated response lacks nuance or detail, it might indicate that the original prompt was too vague or didn't provide enough context. Similarly, if the tone of the AI's response feels overly formal or robotic, this suggests a need to adjust the prompt to elicit a more engaging and relatable answer. These evaluations help educators refine their prompts to better align with teaching goals and create outputs that resonate more effectively with students.

Peer feedback is equally critical in the iterative process of prompt refinement. Constructive criticism should be embraced as a valuable opportunity for growth. Colleagues can provide fresh perspectives, helping to identify strengths, weaknesses, and common patterns in your prompts that you might not have noticed on your own. For example, feedback might reveal that a prompt is overly specific, limiting the AI's ability to generate creative responses, or too broad, leading to generic answers. Reflecting on this input and iterating based on the suggestions strengthens not only the individual prompts but also the educator's overall skill in crafting them.

An effective feedback process requires intentionality. First, when presenting prompts and AI-generated responses for review, clarity and specificity are essential. Vague prompts can lead to ambiguous outputs, while overly detailed prompts might stifle the AI's ability to produce comprehensive insights. Striking a balance ensures that peers can focus on providing actionable feedback instead of deciphering the intent of the prompt. Additionally, pairing positive reinforcement with constructive criticism creates a supportive environment that encourages ongoing collaboration and growth.

The process of analyzing both AI responses and peer feedback fosters better integration of AI tools into the classroom. By identifying areas for improvement, addressing potential biases in responses, and refining prompts through multiple iterations, educators can enhance the quality and effectiveness of AI in their teaching practice. This iterative approach not only improves individual prompts but also builds an educator's confidence and proficiency in leveraging AI as a dynamic educational tool. Furthermore, maintaining empathy and support throughout this process strengthens professional relationships, ensuring continuous improvement and collective success in the use of AI technology in education.

Practical Exercises in Feedback Analysis

To truly excel in feedback analysis for AI prompt engineering, hands-on practice is indispensable. While understanding the theory behind effective prompt crafting is important, the real growth happens when educators actively engage in refining their skills. By analyzing responses, applying peer feedback, and iteratively improving prompts, educators can unlock the full potential of AI to enhance teaching and learning. This process not only improves the quality of AI-generated outputs but also fosters creativity, collaboration, and a deeper understanding of how to align AI tools with educational objectives. The following exercises provide practical, actionable steps for educators to develop and refine these crucial skills.

1. *Prompt Swap and Review*: Pair with a colleague, exchange AI-generated prompts, and provide constructive feedback. Focus on how well the prompts address the educational objective and suggest refinements.
2. *Revise and Resubmit*: Take a previously used prompt, analyze its AI-generated response, and revise it for clarity and specificity. Compare the original and revised outputs to assess improvement.
3. *Feedback Reflection Journal*: Maintain a journal to document feedback received on prompts and the changes made in

response. Reflect on how these adjustments improved outcomes.

4. *Group Discussion on AI Outputs*: Host a group discussion where educators share AI-generated content, collectively analyze the effectiveness of the prompts, and brainstorm improvements.
5. *Case Study Analysis*: Review case studies of AI prompts and responses, identifying strengths and weaknesses and discussing how to refine the prompts for better results.

By engaging in these activities, educators can cultivate a deeper understanding of how feedback shapes the iterative process of AI prompt engineering.

Try it yourself: Feedback Analysis

Experiment with the five exercises above to practice refining your prompts. Identify which methods work best for your feedback community and integrate them into your routine.

Turning Practice Into Mastery

Mastering feedback analysis is not a one-time achievement but an ongoing journey of learning and refinement. These exercises offer educators a structured way to develop their skills, emphasizing collaboration, reflection, and continuous improvement. As educators grow more adept at crafting and refining prompts, they can create AI interactions that are not only more effective but also more engaging and tailored to their students' needs. Through consistent practice, educators can harness AI's potential to transform teaching and learning, fostering a dynamic classroom environment where technology serves as a true partner in education.

Case Study:

Discovering the Power of Feedback in Refining Prompts

I. **Background:** Ms. Carter, a high school science teacher, noticed that her students were disengaged during discussions on renewable energy. She decided to use AI to create a classroom activity that would spark interest and critical thinking. Her initial prompt was, "Explain renewable energy and its benefits." The AI response was accurate but generic, failing to capture the depth and creativity she wanted.

II. **Feedback Process:**

   A. Ms. Carter shared the prompt and the AI's response with a colleague experienced in instructional design.

   B. Feedback indicated the need for specificity and relevance to her student demographic.

   C. Based on this, she revised the prompt to:

> Describe three renewable energy sources and propose a creative way to implement one in a local community project for a 10th-grade science class.

III. **Results:**

A. The revised prompt generated a more detailed and engaging response, including actionable ideas that Ms. Carter used to design a group project.

B. Students were more engaged, discussing how renewable energy could be applied in their community, sparking critical and creative thinking.

IV. **Key Takeaways:**

A. Feedback identified the need for a more specific and student-centered prompt.

B. Iterative revisions improved the relevance and depth of the AI-generated response.

C. Collaborative feedback led to a more impactful classroom activity.

Feedback is a cornerstone of effective AI prompt engineering. By embracing constructive criticism and engaging in reflective practice, educators can refine their prompts to better serve their teaching goals and enhance student engagement. The iterative nature of this process not only improves the quality of AI interactions but also fosters a culture of collaboration and continuous learning. As you integrate feedback into your prompt-crafting practices, you'll find that even small adjustments can lead to significant improvements in educational outcomes.

Chapter 10: Staying Informed and Practicing

In today's fast-paced educational landscape, staying informed about AI is no longer a luxury—it's a necessity. The rapid evolution of AI technology demands that educators adapt continuously to ensure their teaching remains relevant, effective, and engaging. As AI becomes increasingly integral to classrooms, staying updated isn't just about enhancing your own skills; it's about preparing students to thrive in a world where AI plays a significant role in shaping industries and innovation.

Whether you're an experienced educator or just beginning your journey with AI, embracing this dynamic field can feel daunting yet rewarding. By proactively learning and applying the latest AI insights, you ensure that both you and your students stay ahead of the curve. This chapter provides actionable strategies to help you navigate this evolving landscape, integrate AI advancements effectively, and refine your craft as a prompt engineer.

Why Staying Updated Matters

AI's rapid development directly impacts its capabilities and, consequently, the effectiveness of your prompts. Falling behind means missing out on tools and techniques that could enhance your teaching methods and engage students in meaningful ways. As educators, it's crucial to view staying informed as part of the ongoing commitment to lifelong learning. Much like updating lesson plans or adopting new pedagogical methods, integrating the latest AI practices ensures your teaching stays fresh and impactful.

Staying current also helps you understand the nuanced capabilities of AI tools, enabling you to leverage them fully. For example, understanding advancements in natural language processing can help you craft more precise prompts, while familiarity with adaptive AI systems can inform how you personalize learning experiences for your students. Keeping pace ensures that you're not just using AI but using it well.

Case Study:

Adapting to Personalized Learning with AI

I. **Challenge:** A high school science teacher realized that many students were disengaged due to the one-size-fits-all approach to assignments. They wanted to explore how AI could make learning more personalized and engaging.

II. **Approach:**

A. The teacher joined an AI educator forum and learned about AI tools for adaptive learning.

B. They started small, using an AI tool to create personalized study guides based on quiz results.

C. Feedback from students revealed the need for clearer instructions in AI-generated materials, leading the teacher to refine their prompts for better alignment with student needs.

D. Collaboration with colleagues provided additional insights into balancing technology integration with traditional teaching methods.

III. **Outcomes:**

A. Students reported increased engagement and a better understanding of the material.

B. The teacher became more confident in crafting precise prompts that enhanced AI output quality.

C. The experience inspired a school-wide discussion on integrating AI into other subjects.

IV. **Takeaways:**

A. Start small and expand as you gain confidence.

B. Collaborate with peers to share insights and strategies.

C. Use student feedback to refine both AI tools and teaching methods.

Strategies for Staying Informed

As educators, staying informed about advancements in AI is more than a professional responsibility—it's a gateway to transforming your teaching practices and preparing students for a world where technology and learning intersect like never before. The fast-paced nature of AI development means that what worked yesterday may need to be re-evaluated tomorrow. Remaining proactive in understanding these changes ensures that you're not only keeping up but also leveraging AI's capabilities to create dynamic, engaging, and effective classrooms. By committing to continuous learning and skill enhancement, educators can position themselves as leaders in integrating AI thoughtfully and purposefully into education.

Five Key Strategies to Guide the Journey

1. Regular Practice of Prompt Engineering
2. Staying Ahead with AI Developments
3. Engaging with Professional Communities
4. Adapting to New Advancements
5. Embracing an Iterative Learning Process

These steps provide a comprehensive approach to mastering AI's integration into your teaching. From honing your prompt-crafting skills to connecting with a community of like-minded educators, each strategy equips you with the tools to navigate the ever-changing AI landscape effectively. By embedding these practices into your routine, you'll not only improve your proficiency but also ensure that your teaching evolves alongside technology.

The journey to mastering AI in education doesn't end with understanding its current capabilities; it evolves as AI does. By committing to these strategies, you're not only enhancing your own skills but also modeling the importance of adaptability and lifelong learning for your students. This proactive approach ensures that as the world of AI advances, your classroom remains a space where innovation thrives, and students are equipped to navigate and contribute to a tech-driven future. With these tools at your disposal, the opportunities for growth and transformation in your teaching are limitless.

Regular Practice of Prompt Engineering

Incorporating regular practice into your routine is like building muscle memory—it ensures consistent improvement. Set aside time each week to create and refine AI prompts, experimenting across subjects or topics to understand the nuances of effective prompt engineering. For example, one week might focus on crafting prompts for analyzing art, while the next could explore generating writing exercises for English class. This habit keeps your skills sharp and helps you discover how AI can best serve your classroom. To make it manageable, start small—10 minutes twice a week—and increase as it becomes second nature. The goal is to make AI prompt engineering a sustainable part of your teaching practice.

Staying Ahead with AI Developments

Keeping pace with AI advancements is crucial for maintaining a cutting-edge classroom. Dive into AI-focused resources like blogs, webinars, or newsletters from reputable organizations. Join educator forums and professional communities dedicated to integrating AI into teaching. These platforms offer insights into emerging technologies, practical applications, and success stories from fellow educators. For instance, following AI in education blogs can alert you to new tools for personalized learning or classroom management. Staying informed not only enhances your teaching strategies but also ensures that your students benefit from the latest innovations in education technology.

Engaging with Professional Communities

Professional communities offer invaluable support and shared knowledge for navigating the integration of AI in education. Collaborate with colleagues locally or globally through online groups or educator forums. These spaces allow you to exchange ideas, share experiences, and brainstorm solutions to common challenges. For example, you might learn about a new AI tool from a teacher across the country or get feedback on a prompt you've been struggling to refine. Engaging with these communities fosters a culture of continuous learning and collaboration, helping you stay inspired and innovative.

Adapting to New Advancements

The ability to adapt to new advancements is key to leveraging AI effectively. Start by piloting AI tools in small portions of your curriculum, such as using AI for grading or personalized feedback. Gradually expand as you become more comfortable and as the tools prove effective. Collaboration with peers can also enhance adaptation, allowing educators to share strategies and insights. Balancing innovation with practicality ensures that AI tools enhance the learning experience without overwhelming your teaching methods. This iterative approach to adoption keeps your teaching relevant while respecting the established routines that work for you and your students.

Embracing an Iterative Learning Process

AI is a tool, not a replacement for educators. Its role is to enhance teaching, not supplant human expertise. Embracing an iterative learning process means continually updating your skills and knowledge as AI evolves. This involves engaging in professional development opportunities, studying case studies of successful AI integration, and reflecting on your own experiences. For instance, a case study on AI-driven adaptive learning tools might inspire new ways to differentiate instruction in your classroom. By committing to ongoing learning, you ensure that your teaching stays effective and forward-thinking.

Try it yourself: Building Your AI Toolkit

To put these strategies into action, try the following exercise:

- **Develop a Weekly Practice Plan:** Set aside time each week to experiment with crafting and refining AI prompts.
- **Join an AI Educator Forum:** Participate in discussions, share experiences, and learn from peers.
- **Pilot an AI Tool:** Choose one aspect of your curriculum where AI could add value and test it, reflect on the impact.
- **Document and Reflect:** Keep a journal of your AI practices, including prompts you've tried, feedback you've received, and adjustments you've made.

Shaping the Future of Teaching with AI

Staying informed and practicing regularly are the foundation of successful AI integration in education. As educators, we're navigating a transformative era where AI has the potential to enhance, not replace, the critical role we play in shaping young minds. By recognizing that AI tools are here to assist rather than supplant human expertise, we can embrace this shift with confidence and curiosity. Continuous learning, adaptability, and collaboration are key to unlocking AI's potential. Through regular practice of prompt engineering, staying ahead of technological advancements, and engaging with professional communities, we can refine our approaches to teaching and empower students to thrive in a rapidly evolving, AI-driven world.

The integration of AI into teaching marks a pivotal shift in education, one that calls for openness not only to new technologies but also to the psychological and emotional changes that accompany innovation. Professional development becomes essential in equipping educators with the skills and knowledge needed to navigate the complexities of AI tools and pedagogical strategies. Through adaptive learning environments, personalized instruction, and real-world applications, AI offers unprecedented opportunities to enhance accessibility and effectiveness in the classroom. By learning from case studies and collaborating with peers, educators can draw inspiration and gain practical insights into overcoming challenges and leveraging AI's benefits. These experiences underscore the importance of iterative learning processes—of refining, experimenting, and continuously improving.

Educators have a unique role in shaping the future of learning. By embracing ongoing professional growth and adopting a proactive approach to AI, we can create dynamic, student-centered classrooms that prepare learners for the challenges ahead. The journey with AI is not static—it's an evolving process that requires us to innovate, adapt, and share our experiences. Together, we can inspire one another, reimagine teaching, and make a lasting impact on education.

As you close this book, remember that this journey with AI is only beginning. Take bold steps into the world of AI-enhanced teaching, knowing that your creativity, dedication, and expertise are the most important tools in your arsenal. Continue crafting and refining prompts, learning from your peers, and staying curious about the possibilities ahead. Your commitment to growth and innovation will not only transform your classroom but also inspire a new generation of learners. Together, let's shape the future of education—one prompt, one lesson, and one student at a time.

Appendix A: List of Case Studies

Appendix B: Try it yourself

Research

Abdullahi, A. (2023, September 21). *What are large language models?* eWEEK. https://www.eweek.com/artificial-intelligence/large-language-model/

ACU Online Writing Center. (2022, July 15). *EdD in Progress: Hypothetical Case Studies* | ACU Online Writing Center. YouTube. Retrieved January 3, 2024, from https://youtu.be/LYcWypzzqV8?si=la8gnnMoXk2ryOma.

Adams, R. (2023, June 7). *60+ ChatGPT Prompts for Summarizing & Summaries.* AI Habit. https://aihabit.net/chatgpt-prompts-for-summaries/

Adams, R. (2023, November 30). *Crafting the Perfect ChatGPT Prompt: Tips and Tricks (w/ 70+ Prompts)* - AI Habit. AI Habit. https://aihabit.net/crafting-the-perfect-chatgpt-prompt/

Adlawan, D. (2023, October 31). *The Pros and Cons of AI in Education and How it Will Impact Teachers in 2023* | ClassPoint. https://www.classpoint.io/blog/the-pros-and-cons-of-ai-in-education

Advanced discussion leading. (2021, June 30). *Poorvu Center for Teaching and Learning.* https://poorvucenter.yale.edu/teaching/ideas-teaching/advanced-discussion-leading

AiBizBox. (2023, December 16). *Beginner mistakes to avoid when writing prompts.* Ai Biz Box. https://aibizbox.com/ai-prompt-writing/beginner-mistakes-to-avoid-when-writing-prompts/

AI in the Classroom. (2023). Inspiring Inquiry. https://www.inspiringinquiry.com/learningteaching/technology/ai-in-the-classroom

Akgun, S., & Greenhow, C. (2022). *Artificial intelligence in education: Addressing ethical challenges in K-12 settings.* AI and ethics, 2(3), 431–440. https://doi.org/10.1007/s43681-021-00096-7

Akselrod, O. (2023, July 3). *How Artificial intelligence can Deepen racial and economic inequities.* ACLU. American Civil Liberties Union. https://www.aclu.org/news/privacy-technology/how-artificial-intelligence-can-deepen-racial-and-economic-inequities

Alsis, S. (2023, April 4). *Biased AI or: how I learned to stop worrying and love the prompt.* https://www.linkedin.com/pulse/biased-ai-how-i-learned-stop-worrying-love-prompt-sergio-alsis

AltexSoft. (2023, September 13). *Prompt Engineering: The Guide to mastering the art of talking to AI.* AltexSoft. https://www.altexsoft.com/blog/prompt-engineering/

Arkhurst V. (2023, July 20). *Security Risks, bias, AI prompt engineering.* https://www.linkedin.com/pulse/security-risks-bias-ai-prompt-engineering-victoria

Art of AI Prompt Crafting: A Comprehensive guide for Enthusiasts, The. (2023, November 11). Prompting. OpenAI Developer Forum. https://community.openai.com/t/the-art-of-ai-prompt-crafting-a-comprehensive-guide-for-enthusiasts/495144

Artificial Intelligence. (2023, July 7). Office of Educational Technology. https://tech.ed.gov/ai/

Artificial Intelligence Now: ChatGPT & AI Literacy Toolbox: Prompt Literacy. (2024, January 3). FIU Libraries . Florida International University. https://library.fiu.edu/ai/prompt-literacy

Azcuy, R. (2019, August 22). *Engaging students through relevant context.* Teach & Kids Learn (TKL). https://www.teachnkidslearn.com/engaging-students-context/

B12. (2023). *Creating compelling narratives with AI storytelling*. B12. https://www.b12.io/resource-center/ai-how-to-guides/creating-compelling-narratives-with-ai-storytelling.html

Bailey, J. (2023, December 20). *AI in Education.* Education Next. https://www.educationnext.org/a-i-in-education-leap-into-new-era-machine-intelligence-carries-risks-challenges-promises/

Balani, N. (2023, April 23). *Ethical Prompt Engineering: A pathway to responsible AI usage.* Current and Future Technology Trends by Navveen Balani. https://navveenbalani.dev/index.php/articles/ethical-prompt-engineering-a-pathway-to-responsible-ai-usage/

Bansal, S. (2023, November 6). *Top 15 Real world applications of Artificial intelligence.* Uses of AI. Blogs & Updates on Data Science, Business Analytics, AI Machine Learning. https://www.analytixlabs.co.in/blog/applications-of-artificial-intelligence/

Bey, H. H. Y. (2023). *The Benefits of Using AI in Education.* LinkedIn. [Link](https://www.linkedin.com/pulse/benefits-using-ai-education-hakim-h-y-bey-4kkcc)

boost.ai. (2023, November 28). *Six challenges in NLP and NLU - and how boost.ai solves them.* Conversational AI Platform for Enterprise. https://boost.ai/blog/six-nlu-nlp-challenges/

Birss, D. (2023, March 15). *Using AI to view things from multiple perspectives - How to Research and Write Using Generative AI Tools.* [Video]. LinkedIn. https://www.linkedin.com/learning/how-to-research-and-write-using-generative-ai-tools/using-ai-to-view-things-from-multiple-perspectives

Breisacher, J.. (2024, January 16). *Diffit for Teachers: Differentiation with an AI Tool.* Student-Centered World: A Student-Centered Learning Resource. https://www.studentcenteredworld.com/diffit/

Brown, J. (2023, May 15). *Comparing AI Detection Tools: One Instructor's Experience.* The Institute for Learning and Teaching. https://tilt.colostate.edu/comparing-ai-detection-tools-one-instructors-experience/

Burton, C. (2023, July 27). *ChatGPT for Authors: 30+ prompts to help write your next book.* Thinkific. https://www.thinkific.com/blog/ai-to-write-a-book/

Cardon, P. W., Fleischmann, C., Aritz, J., Logemann, M., & Heidewald, J. (2023). *The Challenges and Opportunities of AI-Assisted Writing: Developing AI Literacy for the AI Age.* Business and Professional Communication Quarterly, 86(3), 257–295. https://doi.org/10.1177/23294906231176517

Caruana, Vincent. (2023, November 1). *Top examples of some of the best large language models out there.* Algolia. https://www.algolia.com/blog/ai/examples-of-best-large-language-models/

Casey, K. (2019, January 29). *AI bias: 9 questions leaders should ask.* The Enterprisers Project. https://enterprisersproject.com/article/2019/1/ai-bias-9-questions-leaders-should-ask

Castelvecchi, D. (2023, December 19). AI beats humans on unsolved math problem. Nature Magazine. Scientific American. https://www.scientificamerican.com/article/ai-beats-humans-on-unsolved-math-problem/

Çelik, İ., Dindar, M., Muukkonen, H., & Järvelä, S. (2022). *The Promises and Challenges of Artificial Intelligence for Teachers: a Systematic Review of Research.* TechTrends, 66(4), 616–630. https://doi.org/10.1007/s11528-022-00715-y

Center for Teaching and Learning (2023, June 19). *Pedagogic strategies for adapting to generative AI chatbots.* Stanford. Teaching Commons. https://teachingcommons.stanford.edu/news/pedagogic-strategies-adapting-generative-ai-chatbots

Chacko, A. (2023, November 24). *How to use AI writing prompts to get the best out of your AI tools.* Sprout Social. https://sproutsocial.com/insights/ai-prompt/

Chaudhary, Abhishek. (2023, November 6). *A gentle introduction to large language models.* Towards AI. https://towardsai.net/p/machine-learning/a-gentle-introduction-to-large-language-models

Chen, X., Xie, H., Zou, D., & Hwang, G. (2020). *Application and theory gaps during the rise of Artificial Intelligence in Education.* Computers & Education: Artificial Intelligence, 1, 100002. https://doi.org/10.1016/j.caeai.2020.100002

Chen, Z. (2023, September 13) *Ethics and discrimination in artificial intelligence-enabled recruitment practices. Humanit Soc Sci Commun* **10**, 567. https://doi.org/10.1057/s41599-023-02079-x

Chiancone, C. (2023, July 6). *AI: The Future of Education Harnessing AI to Empower students.* https://www.linkedin.com/pulse/ai-future-education-harnessing-empower-students-chris-chiancone

Chowdhury, M. (2021, December 28). *10 Major challenges of using natural language Processing.* Analytics Insight. https://www.analyticsinsight.net/10-major-challenges-of-using-natural-language-processing/

Clark, H. (2023, November 9). *Collaborating with AI: a modern approach to storytelling in the classroom.* Book Creator App. https://bookcreator.com/2023/08/collaborating-with-ai-a-modern-approach-to-storytelling-in-the-classroom/

ClickUp. (2024) *ChatGPT Prompts for data Analysis.* ClickUpTM. https://clickup.com/templates/ai-prompts/data-analysis

Cooper, K. (2023, December 1). *AI Prompts 101: Understanding How They're Created & Used.* Springboard Blog. https://www.springboard.com/blog/data-science/ai-prompts/

Create Prompts with Clarity and Specificity. (2023, November 25). Pream.ai. https://pream.ai/create-prompts-with-clarity-and-specificity/

Data Incubator, The. (2023, July 12). *AI prompt examples for data scientists to use in 2023.* The Data Incubator. https://www.thedataincubator.com/blog/2023/07/12/ai-in-data-science-ai-prompts-revolutionizing-field/

DeepLearning.AI. (2023, January 11). *Natural Language Processing (NLP) [A complete guide].* https://www.deeplearning.ai/resources/natural-language-processing/

Dieruf, D. (2023, December 8). *What is Prompt Engineering? Everything You Need to Know.* DataStax. https://www.datastax.com/guides/what-is-prompt-engineering

Education Next. (2023*). AI in Education: Leap into a New Era of Machine Intelligence.* Retrieved from https://www.educationnext.org/a-i-in-education-leap-into-new-era-machine-intelligence-carries-risks-challenges-promises/

Ezat, A. (2023, October 24.). *5 Tips To Write Effective Prompts for AI image generation.* Katteb. https://katteb.com/en/tips-to-write-effective-prompts-for-ai-image-generation/

Fard, A. & Oladigbolu, I. (2023, October 24). Why designers should learn prompt Engineering. UI UX Design Agency for SaaS, Fintech & AI | Adam Fard UX Studio. https://adamfard.com/blog/prompt-engineering-for-designers

Febriani, A. (2023, December 25). *Enhancing Higher Order Thinking Skills with AI Integration.* ClassPoint. https://www.classpoint.io/blog/enhancing-higher-order-thinking-skills-with-ai-integration

Ferlazzo, L. (2023, December 31). *I asked artificial intelligence to recommend student writing prompts that can't be easily helped by AI – here are 14 of them.* Larry Ferlazzo's Websites of the Day. https://larryferlazzo.edublogs.org/2023/06/08/i-asked-artificial-intelligence-to-recommend-student-writing-prompts-that-cant-be-easily-helped-by-ai-here-are-14-of-them/

Ferrao, L. (2023, May 8). *How teachers can stay relevant in the age of artificial intelligence?* Leroy Ferrao School. https://www.linkedin.com/pulse/how-teachers-can-stay-relevant-age-artificial-leroy-ferrao-ferrao

Frueh, S. (2023, November 6) *How AI is shaping scientific discovery.* National Academies. https://www.nationalacademies.org/news/2023/11/how-ai-is-shaping-scientific-discovery

Fundamentals, NLP. (2023, October 16). *The 10 Biggest Issues facing Natural Language Processing.* i2. https://i2group.com/articles/the-10-biggest-issues-facing-natural-language-processing

George, A. (2023, June 13). *The importance of artificial intelligence in education for all students.* Language Magazine. https://www.languagemagazine.com/2023/05/31/the-importance-of-artificial-intelligence-in-education-for-all-students/

Gewirtz, D. (2023, November 27). *How to write better ChatGPT prompts for the best generative AI results.* ZDNET. https://www.zdnet.com/article/how-to-write-better-chatgpt-prompts/

Google Developers. (2023). *Introduction to Large Language Models.* [Online]. Available: https://developers.google.com/machine-learning/resources/intro-llms

Govil, A. (2023, August 3). *Unleashing the Power of Prompts with LLM: A Journey to Enhanced Intelligence.* https://www.linkedin.com/pulse/unleashing-power-prompts-llm-journey-enhanced-amit-govil?trk=public_post

Gray Group International. (2023). *How is AI Changing Education?* Gray Group International Blog. [Link](https://www.graygroupintl.com/blog/how-is-ai-changing-education)

Griffy-Brown, C. & Chun, M. (2020, May 5). *Avoiding bias and identifying risks when deploying Artificial Intelligence - a peer-reviewed academic articles: GBR.* Graziadio Business Review | Graziadio School of Business and Management | Pepperdine University. https://gbr.pepperdine.edu/2020/05/avoiding-bias-and-identifying-risks-when-deploying-artificial-intelligence/

Guizeni, S. (2023, June 22). *The Ethical Landscape of Generative AI Prompt Engineering: Examining Unintended Biases and their Social Consequences.* https://www.linkedin.com/pulse/ethical-landscape-generative-ai-prompt-engineering-biases-guizeni

Gupta, R. P. (2023, October 19). *Decoding the magic of prompt engineering in AI - A layman's guide.* https://www.linkedin.com/pulse/decoding-magic-prompt-engineering-ai-laymans-guide-gupta

Gupta, R. P. (2023, November 16). *Reverse engineering in Prompt Engineering: Unraveling AI's creative process.* https://www.linkedin.com/pulse/reverse-engineering-prompt-unraveling-ais-creative-process-gupta-ubxoc

Hernandez, L. (2023, September 13). *The Art of Crafting Inclusive Prompts: A Journey into Diversity in Recruitment.* https://www.linkedin.com/pulse/art-crafting-inclusive-prompts-journey-diversity-larry-hernandez

IBM. (n.d.). *What is Natural Language Processing?* https://www.ibm.com/topics/natural-language-processing

IBM. (n.d.). *What is prompt engineering?* https://www.ibm.com/topics/prompt-engineering

Inclusion Cloud. (2023, June 5). *7 Reasons Prompt Engineering is Essential for Organizations.* Business Insights. https://inclusioncloud.com/insights/blog/prompt-engineering-organizations/

Jennings, J. (2023, December 22). *AI in Education: The Bias Dilemma.* eSpark. https://www.esparklearning.com/blog/get-to-know-ai-the-bias-dilemma/

Johnson, J. (2023, October 26). *The do's and don'ts of writing AI prompts.* Codecademy Blog. https://www.codecademy.com/resources/blog/ai-prompt-engineering-tips/

Jones, B. M. (2023, October 27). *8 Generative AI lessons from the classroom.* Forbes. https://www.forbes.com/sites/beatajones/2023/10/26/8-generative-ai-lessons-from-the-classroom/?sh=6ba57f1b4f52

Kacak, A. (2023, October 5). *ChatGPT, Google Bard, Microsoft Bing, Claude, and Perplexity: Which is the Right AI Tool?* Zeo. https://zeo.org/resources/blog/chatgpt-google-bard-microsoft-bing-claude-and-perplexity-which-is-the-right-ai-to

Karakas, F. (2023, October 25). 100 M*ind-blowing things you need to know about prompt Engineering and how to master prompt engineering for the next decade.* Medium. https://bootcamp.uxdesign.cc/100-mind-blowing-things-you-need-to-know-about-prompt-engineering-and-how-to-master-prompt-5529295f2a9a

Kerner, S. M. (2023, September 13). *large language models (LLMs)*. WhatIs. https://www.techtarget.com/whatis/definition/large-language-model-LLM

Langreo, L. (2023, July 12). *7 Strategies to Prepare Educators to teach with AI.* Education Week. https://www.edweek.org/teaching-learning/7-strategies-to-prepare-educators-to-teach-with-ai/2023/06

Language Magazine. (2023). *The Importance of Artificial Intelligence in Education for All Students.* [Link](https://www.languagemagazine.com/2023/05/31/the-importance-of-artificial-intelligence-in-education-for-all-students/)

Layton, D. (2023, April 10). *Prompt Engineering - leveraging in context learning.* https://www.linkedin.com/pulse/prompt-engineering-leveraging-context-learning-dennis-layton

Levy, M. (2023, May 4). *Your journey in mastering AI communication: How prompt engineering emerges as a crucial skill in the professional world.* https://www.linkedin.com/pulse/your-journey-mastering-ai-communication-how-prompt-engineering-levy

Liu, D. (2023, April 27). *Prompt engineering for educators – making generative AI work for you.* Teaching@Sydney. https://educational-innovation.sydney.edu.au/teaching@sydney/prompt-engineering-for-educators-making-generative-ai-work-for-you/

Lo, L. (2023, August 4). *Human meets AI: Helping educators navigate their emotions about technological change.* EdSurge. https://www.edsurge.com/news/2023-08-04-human-meets-ai-helping-educators-navigate-their-emotions-about-technological-change

Lutkevich, B., & Burns, E. (2023, January 20). *natural language processing (NLP).* Enterprise AI. https://www.techtarget.com/searchenterpriseai/definition/natural-language-processing-NLP

Manyika, J.; Silberg, J.; & Preston, B. (2022, November 17). *What do we do about the biases in AI?* Harvard Business Review. https://hbr.org/2019/10/what-do-we-do-about-the-biases-in-ai

Melo, N. (2023, February 14). *Incorporating Artificial intelligence into the Classroom: An examination of benefits, challenges, and best practices.* eLearning Industry. https://elearningindustry.com/incorporating-artificial-intelligence-into-classroom-examination-benefits-challenges-and-best-practices

McInnes, R. & Kulkarni, A. (2023, October 9). *Mastering generative AI: crafting reusable prompts for effective learning design.* The University of Adelaide. https://www.timeshighereducation.com/campus/mastering-generative-ai-crafting-reusable-prompts-effective-learning-design

McNulty, N. (2023, September 14). Prompt engineering for educational content: Few-Shot and Chain-of-Thought prompting. https://www.linkedin.com/pulse/prompt-engineering-education-content-few-shot-niall-mcnulty

Millner, D. (2023, November 28). *The Future of AI Interaction: The rising importance of prompt Engineering* - HR Curator. HR Curator. https://hrcurator.com/2023/11/28/the-future-of-ai-interaction-the-rising-importance-of-prompt-engineering/

MLA Style Center. (2023, April 12). *How do I cite generative AI in MLA style?* MLA Style Center. https://style.mla.org/citing-generative-ai/

MIT Sloan Teaching & Learning Technologies. (2023, December 18). *Effective Prompts for AI: The Essentials.* MIT Sloan Teaching & Learning Technologies. https://mitsloanedtech.mit.edu/ai/basics/effective-prompts/

Mollick, E. (2023, March 17). *Using AI to make teaching easier & more impactful.* One Useful Thing. https://www.oneusefulthing.org/p/using-ai-to-make-teaching-easier

Mollick, E. (2023, December 7). *An opinionated guide to which AI to use: ChatGPT Anniversary Edition.* One Useful Thing. https://www.oneusefulthing.org/p/an-opinionated-guide-to-which-ai

Mollick, E. & Mollick, L.(2023, October 5). *AI as Feedback Generator.* Harvard Business Publishing. https://hbsp.harvard.edu/inspiring-minds/ai-as-feedback-generator

Mukherjee, S. (2023, June 8). *Prompt Engineering: From Art to Science*. Hexaware -. https://hexaware.com/blogs/prompt-engineering-from-art-to-science-2/

Ng, D. T. K., Luo, W. Y., Chan, H. M. Y., & Chu, S. (2022). *Using digital story writing as a pedagogy to develop AI literacy among primary students.* Computers & Education: Artificial Intelligence, 3, 100054. https://doi.org/10.1016/j.caeai.2022.100054

Nguyen, N. (2023, May 16). *Opportunities and Challenges of AI in Higher Education.* FeedbackFruits. https://feedbackfruits.com/blog/opportunities-and-challenges-of-ai-in-higher-education

Nguyen N. (2023, September 5). *6 ways AI can help save time in facilitating feedback activities.* FeedbackFruits. https://feedbackfruits.com/blog/6-ways-to-save-time-facilitating-feedback-activities-with-ai

Nieves, K. (2023, June 6). *5 ways to use AI tools to meet students' needs.* Edutopia. https://www.edutopia.org/article/using-ai-tools-differentiated-instruction/

Olavsrud, T. B. T. (2023, August 11). *What is NLP? Natural language processing explained.* CIO. https://www.cio.com/article/228501/natural-language-processing-nlp-explained.html

OpenAI. (2024). ChatGPT 3.5 (January 27 version) [Large language model]. https://chat.openai.com/chat

Patil, S. (2023, November 13). *Context window size and language model performance: Balancing Act.* Medium. https://ai.plainenglish.io/context-window-size-and-language-model-performance-balancing-act-2ae2964e3ec1

PixelCrayons. (2023). *The Importance of AI in Education*. PixelCrayons Blog. [Link](https://www.pixelcrayons.com/blog/digital-transformation/importance-of-ai-in-education/)

Pictory. (2023, December 13). *What are AI prompts? examples, optimization guide, and more*. Pictory.ai. https://pictory.ai/blog/what-are-ai-prompts-examples-optimization-guide-and-more-2023?el=0115&htrafficsource=pictoryblog

Poth, R. D. (2023, November 9). *Effective professional development on AI.* Edutopia. https://www.edutopia.org/article/ai-professional-development-helps-teachers-tech-integration/

Prompt Literacy in Academic. (2023). U-M Generative AI. https://genai.umich.edu/resources/prompt-literacy

Qwiklabs. (2023, October 12). *Introduction to large language Models.* Google Cloud Skills Boost. https://www.cloudskillsboost.google/course_templates/539

Ramlochan, S. (2023, April 5). *Master Prompt Engineering: Demystifying prompting through a structured approach.* Prompt Engineering. https://promptengineering.org/master-prompt-engineering-demystifying-prompting-through-a-structured-approach/

Ramlochan, S. (2023, August 6). *Unlock the Power of AI with Effective Prompts.* Prompt Engineering. https://promptengineering.org/master-prompt-engineering-ai-prompt/

Ruiz, P. (2023, November 3). *Artificial Intelligence in Education: A reading guide focused on Promoting Equity and Accountability in AI.* https://circls.org/educatorcircls/ai-in-education/ai-in-ed-reading-guide

Russell, S. (2023, November 2). *Using AI for higher ed marketing?* Beware the AI bias. EducationDynamics. https://www.educationdynamics.com/using-ai-for-higher-ed-marketing-beware-the-ai-bias/

Sachdeva, N. (2023, April 13). *How to tackle bias in AI: An Ultimate Guide.* Daffodil Unthinkable Software Corp. https://insights.daffodilsw.com/blog/ai-bias

Saleem, A. (2023, July 14). *Master prompt engineering with effective strategies.* Data Science Dojo. https://datasciencedojo.com/blog/prompt-engineering-strategies/#

Samarth, V. (2023, December 14). *What is Prompt Engineering, and What is its Significance in Today's World?* Emeritus India. https://emeritus.org/in/learn/ai-and-ml-prompt-engineering/

Sample, I. (2023, December 15). *AI scientists make 'exciting' discovery using chatbots to solve maths problems.* The Guardian. https://www.theguardian.com/science/2023/dec/14/ai-scientists-make-exciting-discovery-using-chatbots-to-solve-maths-problems

Sathiaseelan, A. (2023, August 22). *The Art of Prompt Engineering: Enhancing AI language models for specific tasks*. https://www.linkedin.com/pulse/art-prompt-engineering-enhancing-ai-language-models-sathiaseelan

S*econd proprietary GPT model, free to use for now, that rivals ChatGPT4, A.* Community. IPFire Community. (2023, July 12). IPFire Community. https://community.ipfire.org/t/a-second-propretary-gpt-model-free-to-use-for-now-that-rivals-chatgpt4/10037

Serrels, M. (2023, June 30). *AI's teachable moment: How ChatGPT is transforming the classroom.* CNET. https://www.cnet.com/tech/features/ais-teachable-moment-how-chatgpt-is-transforming-the-classroom/

Shanklin, R., Samorani, M., Harris, S. L., & Santoro, M. A. (2022). *Ethical Redress of Racial Inequities in AI: Lessons from Decoupling Machine Learning from Optimization in Medical Appointment Scheduling.* Philosophy & Technology, 35(4). https://doi.org/10.1007/s13347-022-00590-8

Sharma, R. (2023, August 8). *What is Prompt Engineering, and What is its Significance in Today's World?* Emeritus Online Courses. https://emeritus.org/blog/ai-and-ml-prompt-engineering/

Sharma, S. (2023, August 14). *Guiding students to assess the merits of artificial intelligence tools.* Edutopia. https://www.edutopia.org/article/teaching-students-use-ai-tools/

Signet Education. (2023, July 18). *Artificial vs. Authentic: Four reasons why using AI to write your college essay is a bad idea.* Signet Education. Signet Education - Online Individualized Tutoring & Test Prep | Signet Education. https://signeteducation.com/resources/artificial-vs-authentic-four-reasons-why-using-ai-to-write-your-college-essay-is-a-bad-idea/

Snyder, K. (2023, May 18). *Mindful AI: Crafting prompts to mitigate the bias in generative AI.* Textio. https://textio.com/blog/mindful-ai-crafting-prompts-to-mitigate-the-bias-in-generative-ai/115959775665

Soluntech. (2023, August 14). *How to write effective prompts for AI Systems.* Custom Software Services for Forward-Thinking Organizations. https://www.soluntech.com/blog/how-to-write-effective-prompts-for-ai-systems

Sperduto, B. (2023, October 11). *Keep AI in Check: A Managing Editor's Checklist for Fact-Checking AI Outputs.* Leading Content Marketing Platform |

Contently. https://contently.com/2023/10/11/keep-ai-in-check-a-managing-editors-checklist-for-fact-checking-ai-outputs/

Suess, T. (2023, September 26). *Mastering Prompt Engineering: A Multi-Level Approach to Problem-Solving with AI.* https://www.linkedin.com/pulse/mastering-prompt-engineering-multi-level-approach-ai-timm-suess

Takyar, A. (2023, December 19). *A comprehensive guide to prompt engineering.* LeewayHertz - AI Development Company. https://www.leewayhertz.com/prompt-engineering/

Taskade. (2023) *AI Inclusive Language Guide Generator.* Taskade. https://www.taskade.com/generate/human-resource/inclusive-language-guide

Tobin, J. (2023, November 21). *Educational technology: Digital innovation and AI in schools.* House of Lords Library. https://lordslibrary.parliament.uk/educational-technology-digital-innovation-and-ai-in-schools/

Todoros, O. (2023, October 10). *Unlocking the Future of AI — A Deep Dive into AI Prompts and Their Applications.* Spike. https://www.spikenow.com/blog/productivity/unlocking-the-future-of-ai-a-deep-dive-into-ai-prompts-and-their-applications/

Trackmind. (2023, September 13). *Designing Effective Prompts: A guide to prompt engineering.* https://www.linkedin.com/pulse/designing-effective-prompts-guide-prompt-engineering-trackmind

University of San Diego. (n.d.). *Artificial Intelligence in Education.* Retrieved from https://onlinedegrees.sandiego.edu/artificial-intelligence-education/

Cardona, M., Rodríguez, R, & Ishmael, K. (2023, May) *Artificial Intelligence and Future of Teaching and Learning: Insights and Recommendations.* Office of Educational Technology. Washington, DC. https://www2.ed.gov/documents/ai-report/ai-report.pdf.

Vergadia, P., & Williams, K. (2023, August 14). *Best practices for prompt engineering.* Google Cloud Blog. https://cloud.google.com/blog/products/application-development/five-best-practices-for-prompt-engineering

Victor, A. (2024, January 16). *Prompt Engineering: the key to ethical AI conversations.* https://www.linkedin.com/pulse/prompt-engineering-key-ethical-ai-conversations-adam-m-victor-drqqc?trk=article-ssr-frontend-pulse_more-articles_related-content-card

Vojtaskovic, P. (2023, August 1). *The significance of high quality AI prompts.* https://www.linkedin.com/pulse/significance-high-quality-ai-prompts-pat-vojtaskovic

Weiss, A. (2023, October 4). *How to Differentiate Instruction with AI.* Khan Academy Blog. https://blog.khanacademy.org/how-to-differentiate-instruction-with-ai/

Williamson, B. (2023). *The Social life of AI in Education.* International Journal of Artificial Intelligence in Education. https://doi.org/10.1007/s40593-023-00342-5

Workiva. (2023, November 21). *Everything You Need to Know About Writing Gen AI Prompts*. Workiva. https://www.workiva.com/blog/how-to-write-ai-prompts

Yasar, K. (2023, September 26). *AI prompt.* Enterprise AI. https://www.techtarget.com/searchenterpriseai/definition/AI-prompt

Zhang, K., & Aslan, A. (2021). *AI technologies for education: Recent research & future directions.* Computers & Education: Artificial Intelligence, 2, 100025. https://doi.org/10.1016/j.caeai.2021.100025

Zharovskikh, A. (2023, December 11). *AI prompt engineering benefits*. InData Labs. https://indatalabs.com/blog/ai-prompt-engineering

www.ingramcontent.com/pod-product-compliance
Lightning Source LLC
Chambersburg PA
CBHW061557260225
22526CB00062B/523
* 9 7 8 1 7 3 6 2 7 0 7 5 2 *